Basics of
DIFFERENTIAL GEOMETRY

Basics of
DIFFERENTIAL GEOMETRY

Rohit Garg

ANMOL PUBLICATIONS PVT. LTD.
NEW DELHI-110 002 (INDIA)

ANMOL PUBLICATIONS PVT. LTD.
Regd. Office: 4360/4, Ansari Road, Daryaganj,
New Delhi-110002 (India)
Tel.: 23278000, 23261597, 23286875, 23255577
Fax: 91-11-23280289
Email: anmolpub@gmail.com
Visit us at: www.anmolpublications.com

Branch Office: No. 1015, Ist Main Road, BSK IIIrd Stage
IIIrd Phase, IIIrd Block, Bengaluru-560 085 (India)
Tel.: 080-41723429 • Fax: 080-26723604
Email: anmolpublicationsbangalore@gmail.com

Basics of Differential Geometry

First Edition, 2011

ISBN 978-81-261-4839-4

PRINTED IN INDIA

Printed at Suman Printers, Delhi

Contents

Preface

Differential geometry is a mathematical discipline using the techniques of differential and integral calculus, as well as linear and multilinear algebra, to study problems in geometry. The theory of plane and space curves and of surfaces in the three-dimensional Euclidean space formed the basis for its initial development in the eighteenth and nineteenth centuries. Since the late nineteenth century, differential geometry has grown into a field concerned more generally with geometric structures on differentiable manifolds.

It is closely related to differential topology, and to the geometric aspects of the theory of differential equations. Grigori Perelman's proof of the Poincaré conjecture using the techniques of Ricci flow demonstrated the power of the differential-geometric approach to questions in topology and highlighted the important role played by the analytic methods. Differential geometry of surfaces already captures many of the key ideas and techniques characteristic of the field.

Author

1

Euclidean Differential Geometry

INTRODUCTION

Euclidean geometry, sometimes called parabolic geometry, is divided into two subsections: plane geometry, geometry dealing with figures in a plane, and solid geometry, geometry dealing with solids in three-dimensional space. Solid geometry is sometimes called three-dimensional Euclidean geometry and deals with solids, such as polyhedra and spheres, and lines and planes in three-dimensional space.

Solid geometry is concerned with the study of figures in three-dimensional Euclidean space, which is usually denoted R^3. Solid geometry is a branch of Euclidean geometry, developed by Greek mathematician Euclid in the 4th century B.C., that is governed by Euclid's five postulates as laid out in his work *T*.

Euclid departed from traditional definitions and instead used motion-based definitions, and 39 propositions. These motion-based definitions were formulated as revolutions of figures about a diameter. The definitions cover everything from the definition of a solid, inclination of figures in three-dimensional space, similarity of those figures, definitions of pyramids, spheres, cones, cylinders, cubes, octahedrons, icosahedrons, and dodecahedrons.

There are also several definitions that deal with angles and angular relations of figures relative to each other. The first 18 propositions deal with lines and planes and their relations to figures in Euclidean space. The elementary theorems of three-dimensional geometry, solid geometry, are laid out in this book. The rest of the propositions deal with planes and solids in Euclidean space. Solid geometry is an extension and reinforcement of the propositions describing plane geometry. It forms the basis for the foundations of other branches of mathematics including geometry and trigonometry.

The development of geometry in general was driven by three classical problems in Greek mathematics: plane, solid, and linear problems. Those problems whose solutions employ the use of one or more sections of a cone are solid problems.

GEOMETRY

Geometry was extremely important to ancient societies and was used for surveying, astronomy, navigation, and building. Geometry, as we know it is actually known as Euclidean geometry which was written well over 2000 years ago in Ancient Greece by Euclid, Pythagoras, Thales, Plato and Aristotle just to mention a few. The most fascinating and accurate geometry text was written by Euclid, and was called Elements. Euclid's text has been used for over 2000 years!

Geometry is the study of angles and triangles, perimeter, area and volume. It differs from algebra in that one develops a logical structure where mathematical relationships are proved and applied.

TERMS

1. *Point:* Points show position. A point is shown by one capital letter. In the example below, A, B, and C are all points. Notice that points are on the line.

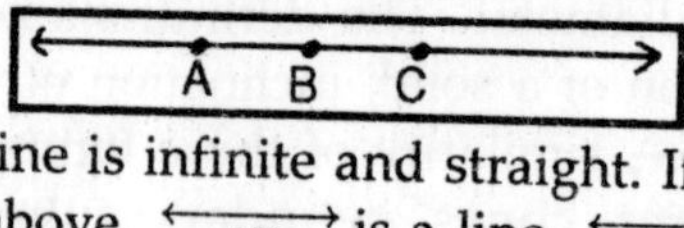

2. *Line:* A line is infinite and straight. If you look at the picture above, $\overleftrightarrow{AB}$ is a line, $\overleftrightarrow{AC}$ is also a line and $\overleftrightarrow{BC}$ is a line. A line is identified when you name two points on the line and draw a line over the letters. A line is a *set* of continuous points that extend indefintely in either of its direction. Lines are also named with lowercase letters or a single loswer case letter.

3. *Line Segment:* A line segment is a straight line segment which is part of the straight line between two points. To identify a line segment, one can write AB or $\overleftrightarrow{AB}$. The points on each side of the line segment are referred to as the end points.
4. *Ray:* A ray is the part of the line which consists of the given point and the set of all points on one side of the end point. $\overleftrightarrow{AB}$ A is the end point and this ray means that all points starting from A are included in in the ray. A ray can also be written like: $\overleftrightarrow{AB}$
5. *Angle:* An angle can be defined as two rays or two line segments having a common end point. The endpoint becomes known as the *vertex*. An *angle* occurs when two rays meet or unite at the same endpoint. The angles pictured below can be identified as $\angle ABC$ or $\angle CBA$. You can also write this angle as $\angle B$ which names the vertex. (common endpoint of the two rays.)

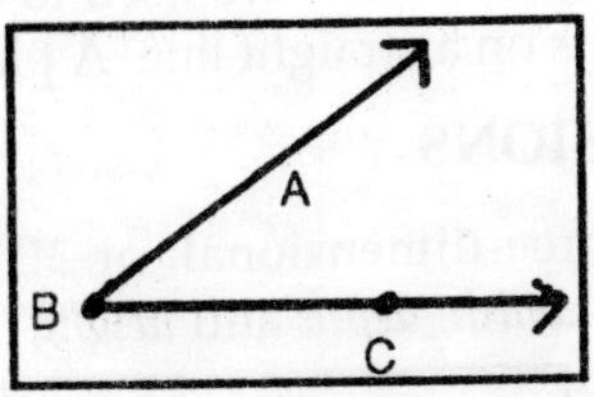

The vertex (in this case B) is always written as the middle letter. It matters not where you place the letter or number of your vertex, it is acceptable to place the it on the inside or the outside of your angle.

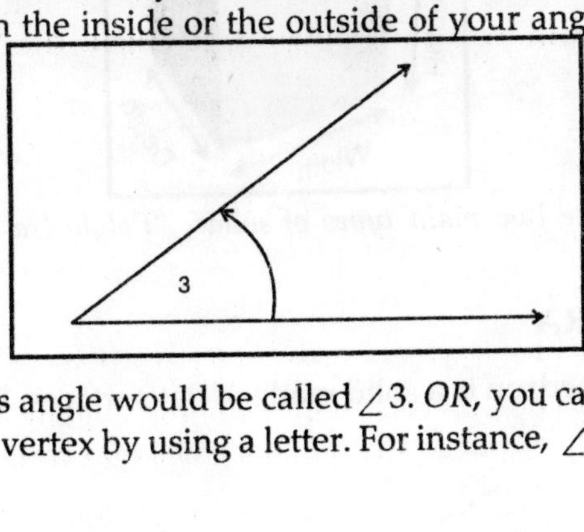

This angle would be called $\angle 3$. *OR*, you can also name the vertex by using a letter. For instance, $\angle 3$ could also be named angle B if you choose to change the number to a letter.

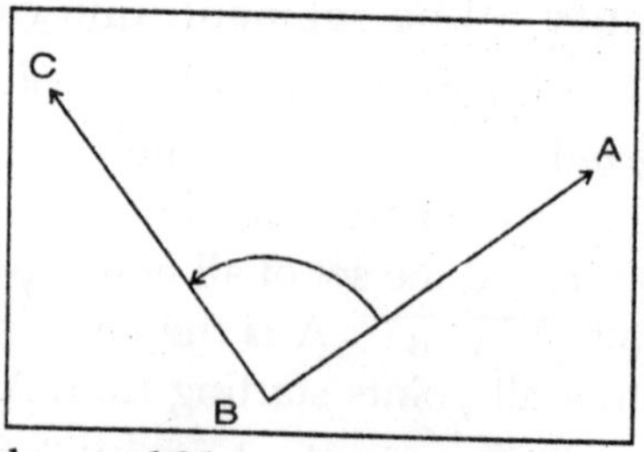

This angle would be named ∠ABC or ∠CBA or ∠B When you are referring to your text book and completing homework, make sure you are consistent! If the angles you refer to in your homework use numbers - use numbers in your answers. Which ever naming convention your text uses is the one you should use.

6. *Plane:* A plane is often represented by ablackboard, bulletin board, a side of a box or the top of a table. These 'plane' surfaces are used to connect any two or more points on a straight line. A plane is a flat surface.

THREE DIMENSIONS

It is called three-dimensional, or 3D because there are three dimensions: *width, depth* and *height.*

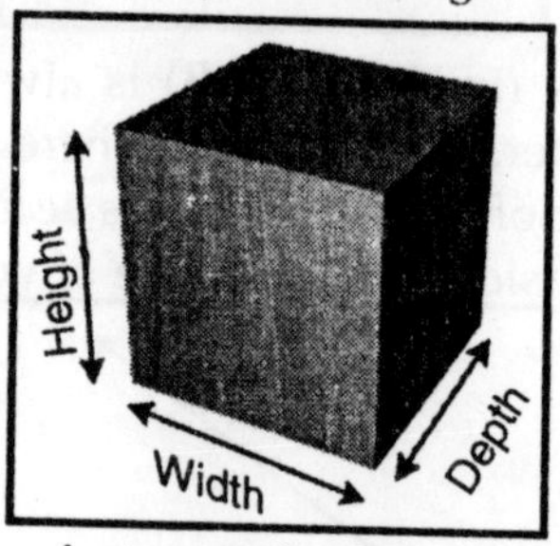

There are two main types of solids, "Polyhedra", and "Non-Polyhedra":

POLYHEDRA

A polyhedron is a solid made of flat surfaces. Each surface is a polygon.

Examples:

Fig. Prisms

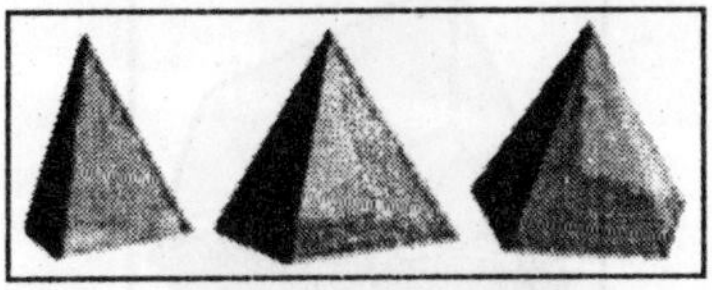

Fig. Pyramids

Fig. Platonic Solids

Platonic Solids

For each solid there are two printable nets. These nets can be printed onto a piece of card. A plane may be measured in terms of two dimensions: length and width. Solid geometry deals with figures that have three dimensions: length, width, and height. Conic sections, a commonly studied topic of geometry, are two-dimensional curves created by slicing a plane through a three-dimensional hollow cone.

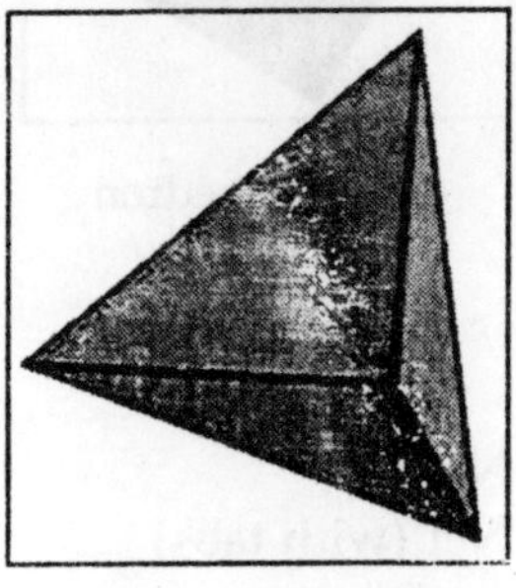

Fig. Tetrahedron

- 4 Faces
- 4 Vertices
- 6 Edges
- Tetrahedron Net
- Tetrahedron Net (with tabs)
- Spin a Tetrahedron

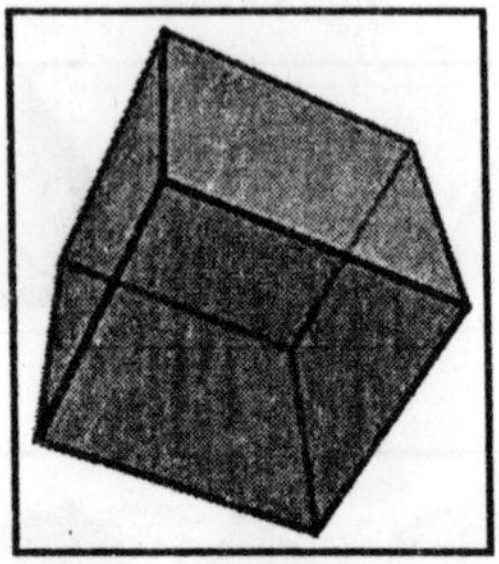

Fig. Cube

- 6 Faces
- 8 Vertices
- 12 Edges
- Cube Net
- Cube Net (with tabs)
- Spin a Cube

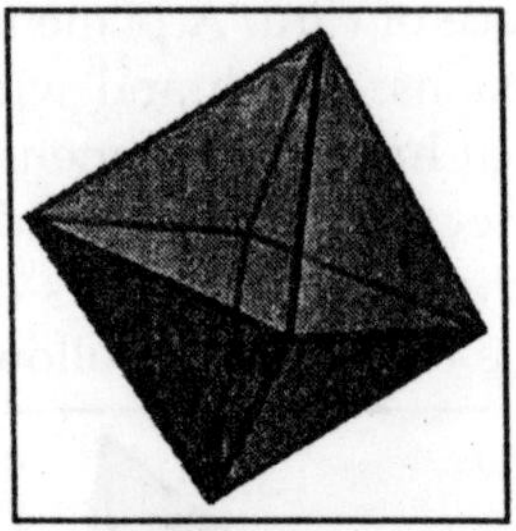

Fig. Octahedron

- 8 Faces
- 6 Vertices
- 12 Edges
- Octahedron Net
- Octahedron Net (with tabs)
- Spin an Octahedron

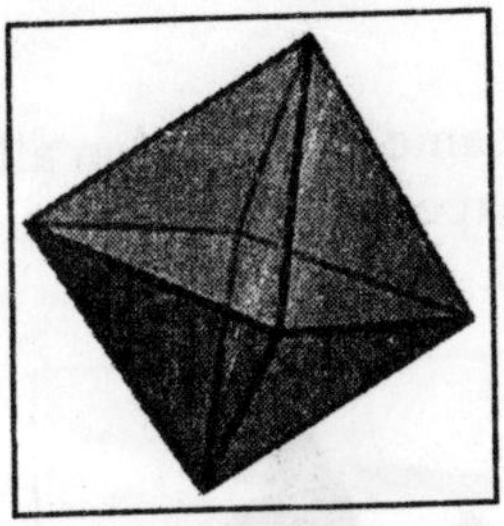

Fig. Dodecahedron

- 12 Faces
- 20 Vertices
- 30 Edges
- Dodecahedron Net
- Dodecahedron Net (with tabs)
- Spin a Dodecahedron

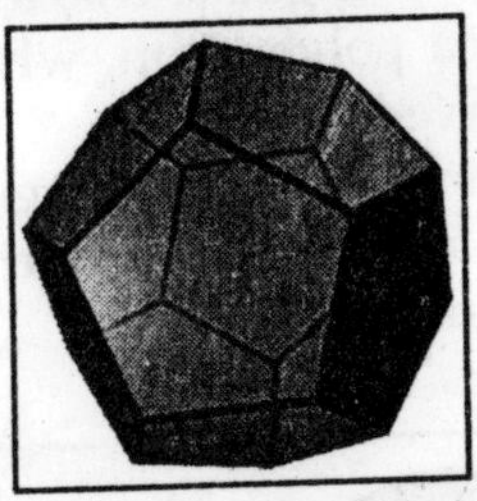

Fig. Icosahedron

- 20 Faces
- 12 Vertices
- 30 Edges
- Icosahedron Net
- Icosahedron Net (with tabs)
- Spin an Icosahedron

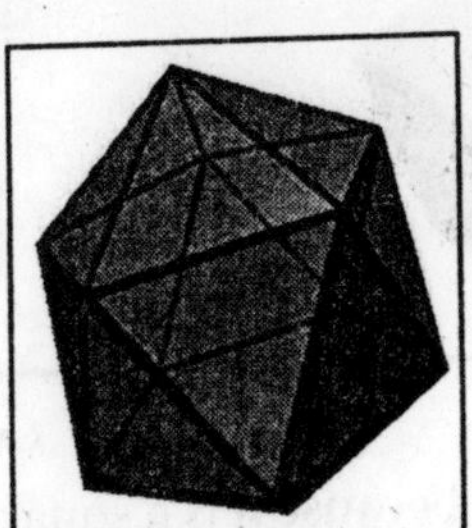

Prisms

A prism has the same cross section all along its length ! A cross section is the shape you get when cutting straight across an object.

The cross section of this object is a triangle it has the same cross section all along its length..and so it's a triangular prism. A prism is officially a polyhedron, which means all sides should be flat. No curved sides.

So the cross section will be a polygon (a straight-edged figure). For example, if the cross section was a circle then it would be a cylinder, not a prism.

These are all Prisms:

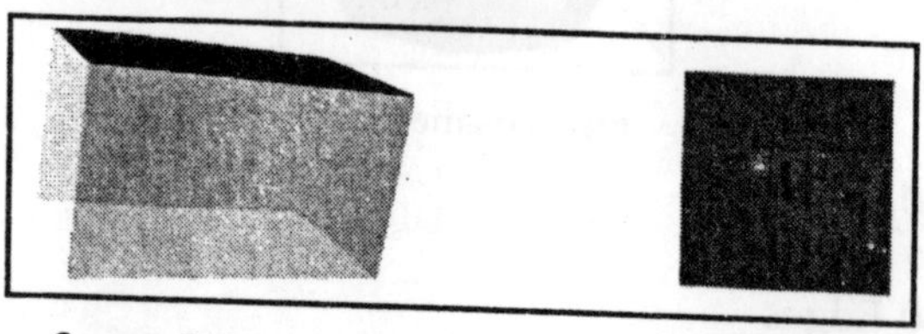

Square Prism: *Cross-Section*:

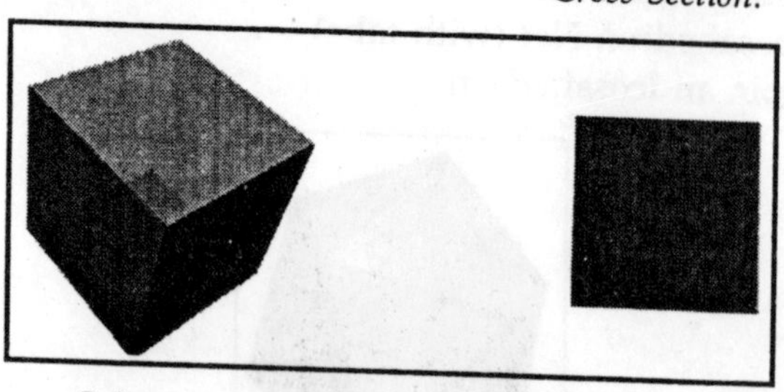

Cube: *Cross-Section*:

(yes, a cube is a prism, because it is a square all along its length)

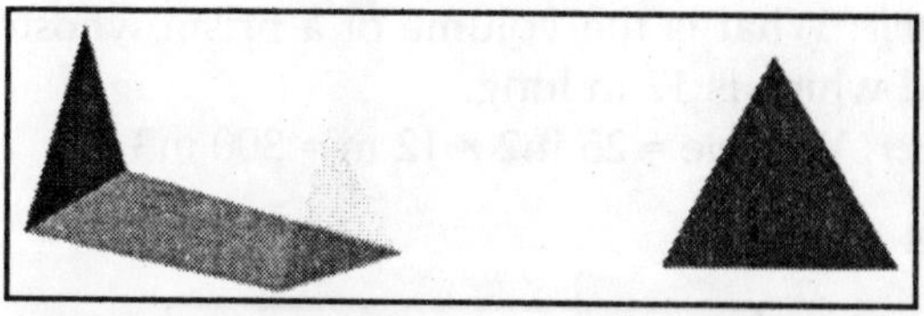

Pentagonal Prism: *Cross-Section:*

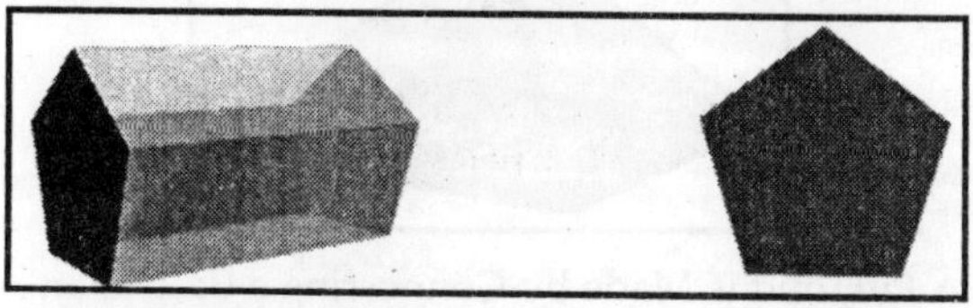

Regular and Irregular Prisms: All the previous examples are Regular Prisms, because the cross section is regular (in other words it is a shape with equal edge lengths)

Here is an example of an Irregular Prism:

Irregular Pentagonal Prism:Cross-Section: (It is "irregular" because the Pentagon is not "regular"in shape)

Volume of a Prism: The Volume of a prism is simply the area of one end times the length of the prism

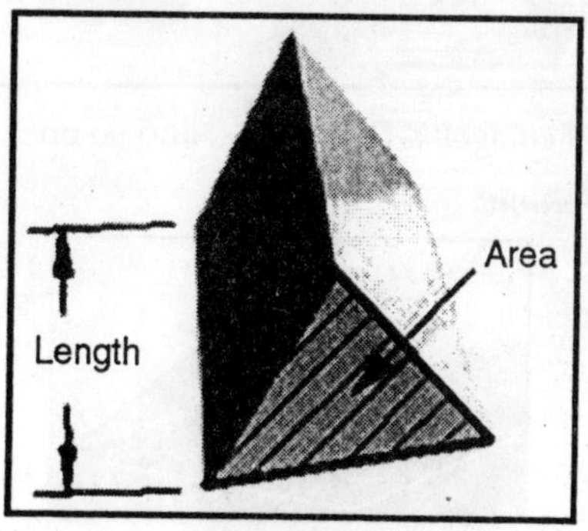

Volume = Area × Length

Example: What is the volume of a prism whose ends are 25 in2 and which is 12 in long.

Answer: Volume = 25 in2 × 12 in = 300 in3

Pyramids

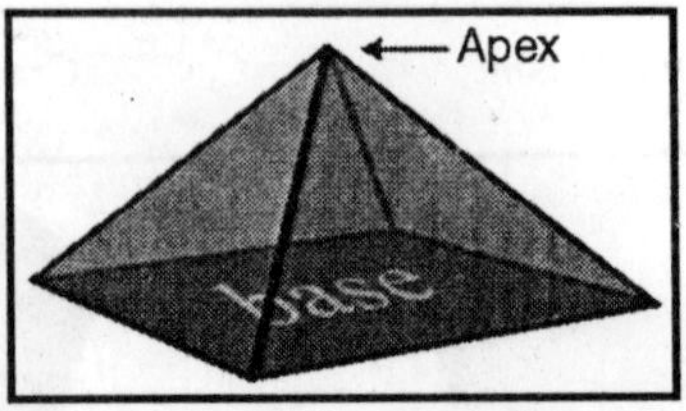

Fig. A Pyramid is Made by Connecting a Base to an Apex

Types of Pyramids: There are many types of Pyramids, and they are named after the shape of their base.

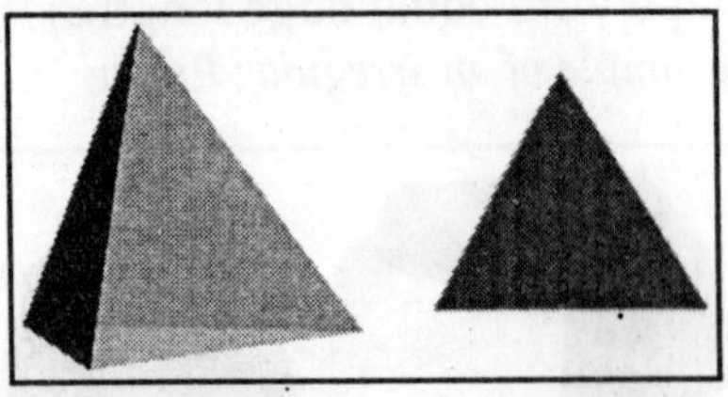

Triangular Pyramid:

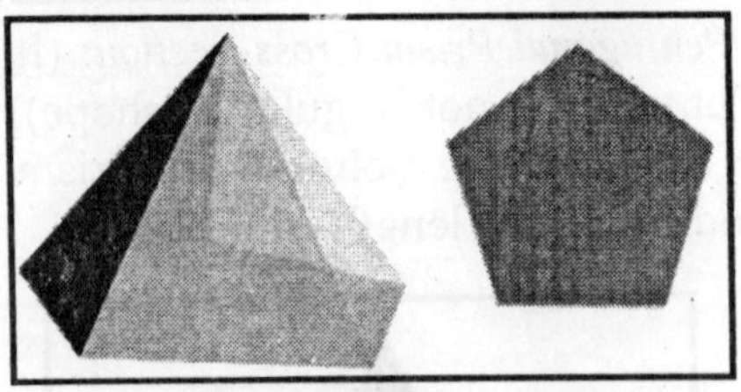

Pentagonal Pyramid: ... and so on...

Area and Volume:

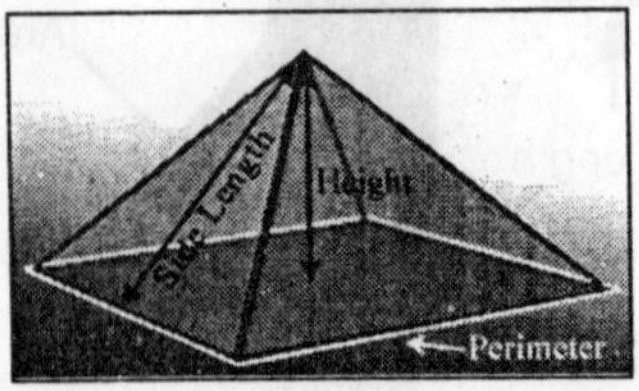

- The Volume of a Pyramid: 1/3 × [Base Area] × Height

- The Surface Area of a Pyramid: 1/2 × Perimeter × [Side Length] + [Base Area]

Explanation of Surface Area: The Surface Area has two parts: the area of the sides (the Lateral Area) and the area of the base (the Base Area). The Base Area depends on the shape, there are different formulas for triangle, square, etc. See Area for formulas, or our Area Calculation Tool. But the Lateral Area is surprisingly simple. Just multiply the perimeter by the side length and divide by 2. This is because the sides are always triangles and the triangle formula is base times height divided by 2

Right vs Oblique Pyramid: This tells you where the top (apex) of the pyramid is. If the apex is directly above the center of the base, then it is a Right Pyramid, otherwise it is an Oblique Pyramid.

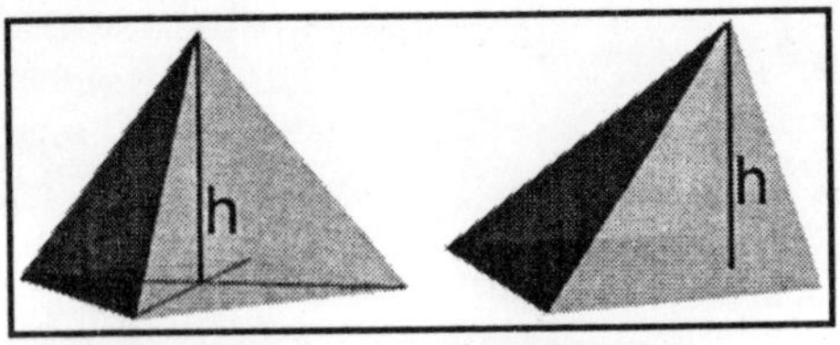

Right Pyramid Oblique Pyramid

Regular vs Irregular Pyramid: This tells us about the shape of the base. If the base is a regular polygon, then it is a Regular Pyramid, otherwise it is an Irregular Pyramid.

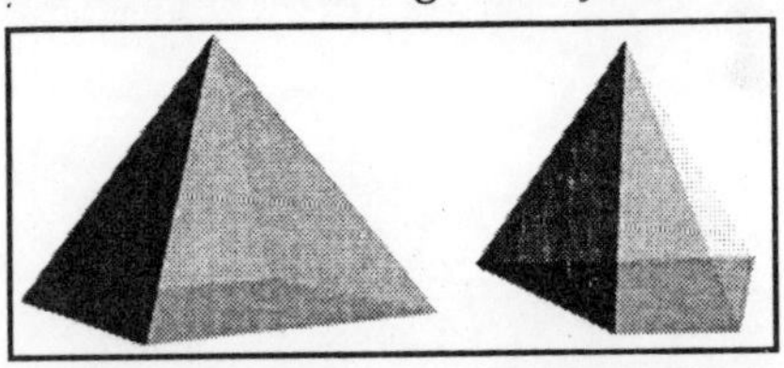

Regular Pyramid Irregular Pyramid

Base is Regular Base is Irregular

NON-POLYHEDRA

Some solids have curved surfaces, or a mix of curved and flat surfaces (so they aren't polyhedra).

Examples:

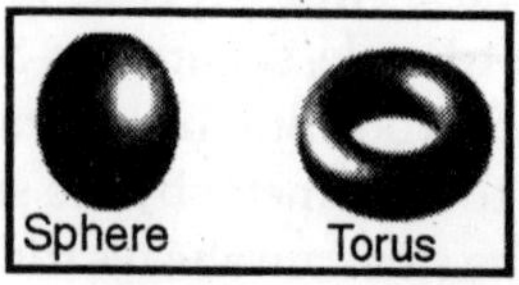

2

Differential Curves

CONICS

Many people have contributed to this study, and many historical references and texts exist to document this study. A section is the surface or outline of that surface formed by cutting a solid figure with a plane. If the solid figure is a cone, the resulting curve is called a conic section. The diagram below shows a cone formed by rotating a diagonal line around a vertical axis. Four planes are shown, cutting through the cone at various angles, producing the curves shown in the following diagram. The intersection of each plane with the cone forms a conic section, with the focus of the conic section at the place where the axis of the cone passes through the section. The kind and shape of the conic section is determined by the angle of intersection of the plane, and the axis of the cone.

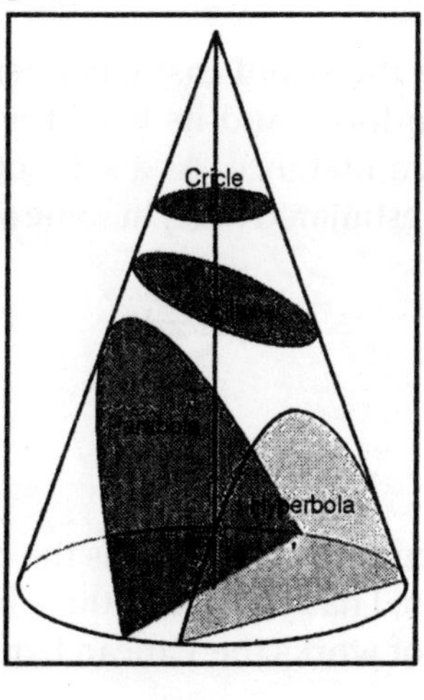

Angled view of a cone, with conic sections produced by cutting the cone at different angles. Cutting at right angles to the axis produces a circle. Cutting at less than a right angle to the axis, but more than the angle made by the side of the cone, produces an ellipse.

Cutting parallel to a side of the cone produces a parabola. Cutting more nearly parallel to the axis than to the side produces a hyperbola. Where the axis cuts each section is a focus of that section.

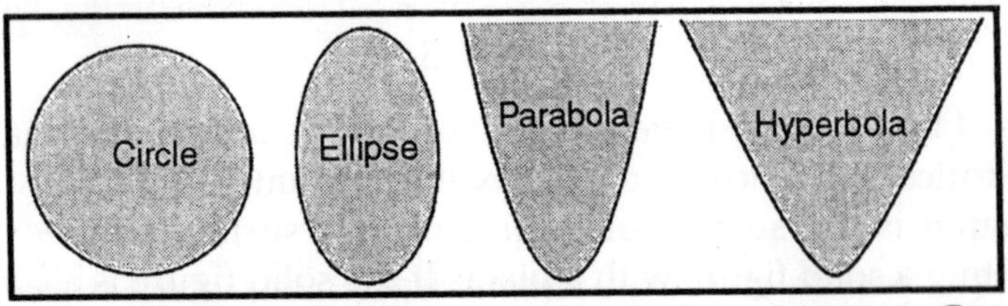

Apollonius of Perga, one of the greatest Greek mathematicians of the time appears to have been the first to have rigorously studied the conic sections. He applied his work to his study of planetary motion and used this to aid in the development of Greek astronomy.

We see that Perga was one of the cities visited by the apostle Paul during his first missionary journey as recorded in Acts 13:13. Paul would have been in Perga less than 300 years after Apollonius' development of the topic of conic sections.

Now he made the sea of cast metal ten cubits from brim to brim, circular in form, and its height was five cubits, and thirty cubits in circumference. As a sidenote, observe that the value of π is being estimated here, in some sense, by the integer 3. That is to say,

$$C = \pi d$$

or

$$\pi = \frac{c}{d}$$

where C is the circumference and d is the diameter of a circle. To close this section, I have included three images below which are reproductions of works the Vatican Exhibit Main Hall. This site contains many old manuscripts of writings in the areas of mathematics and astronomy, as well as concise background material to shed light on the historical significance of the documents.

GEOMETRIC ORIGIN OF CONIC SECTIONS

It was Apollonius who was the first to note that the conic sections could be constructed apart from algebraic equations by cutting the right--circular cone with a plane. As a matter of fact, Apollonius did not note the connection of the conics to their algebraic equations. These equations did not enter the mathematical picture for hundreds of years.

Let us examine the figure above a little more closely. Note that if the right--circular cone is cut by a plane which is perpendicular to the axis of the cone, then the resulting intersection is a circle. (Note that the axis of the cone in the figure is simply the vertical line which goes through the vertex of the cone, the point at which the two pieces of the cone meet.) If the cutting plane is parallel to the ``side'' of one of the cones, then we see that a parabola is constructed. Thirdly, if the plane intersects both pieces of the cone, then a hyperbola appears. Finally, if the plane intersects one of the pieces of the cone but is NOT parallel to the ``side'' of one of the cones, then an ellipse is generated.

ALGEBRAIC VIEW

We now move from a geometric point of view to an algebraic approach. Note that all of the conics that we will see below will satisfy a second-degree equation of the form

$$Ax^2 + Bxy + Cy^2 + Dx + Ey + F = 0$$

where *A, B, C, D, E,* and *F* are constants. As we change.

The values of some of these constants, the shape of the corresponding conic will also change. Hence, it will be very important to focus on these differences in the algebraic equations as we study the individual conic sections. Knowing the subtle differences in the equations will help us to quickly identify the type of conic that is represented by a given equation.

You have undoubtedly already worked with such equations in your previous mathematics courses, although you may not realise it at this point. For example, the equation,

$$x^2 + y^2 - 25 = 0$$

is of the form mentioned above. It can be rewritten as,

$$x^2 + y^2 = 25,$$

which you may recognize as the equation for a circle centered at the origin with radius 5.

So you see, you probably have worked with these algebraic equations before. They will continue to appear as you move into the sections below dealing with the particular conic sections.

KINDS OF CONIC SECTIONS

When the plane cuts the cone at right angles to its axis, a circle is formed. The axis passes through the center of the circle, and the focus (and the Sun, if the curve represents the orbit of a planet) is at the center of the circle.

When the plane cuts the cone parallel to the side of the cone, a parabola is formed. Since parallels never meet, the plane never cuts that side of the cone that it is parallel to, and the arms of the parabola extend into infinity, eventually becoming parallel to each other, and to the axis of the parabola.

Since circles and parabolas are formed by cutting the cone at specific angles, they have unique shapes. All circles are identical in shape, and all parabolas are identical in shape; only their size and orientation differ. For ellipses and hyperbolas, however, there is a wide range of angles between the plane and the axis of the cone, so they have a wide range of shapes.

When the plane cuts the cone at an angle between a perpendicular to the axis (which would produce a circle), and an angle parallel to the side of the cone (which would produce a parabola), the curve formed is an ellipse. Since circles and parabolas are formed by angles just beyond the range of angles which produce ellipses, ellipses can vary in shape from very nearly circular, to very nearly parabolic. The closer the plane is to a perpendicular to the axis, the more nearly circular the ellipse is, and the closer its focus lies to its center. The closer the plane is to being parallel to the side of the cone, the more elongated the ellipse is, the closer its focus is to one end of the ellipse, and the more either end of the ellipse looks similar to a parabola.

When the plane cuts the cone at an angle closer to the axis than the side of the cone, a hyperbola is formed. As in the case of a parabola, the curve extends into infinity, as the plane can never reach the far side of the cone; and in fact, gets further and further from it, the further along the arms you go. Hyperbolas which are formed by angles close to the side of the cone look very nearly parabolic, whereas hyperbolas which are formed at steeper angles look less parabolic; but in every case, there is a fundamental difference between a hyperbola and a parabola -- the arms of a parabola eventually become parallel to each other, while the arms of a hyperbola always make an angle relative to each other. For a hyperbola which is very nearly parabolic, this angle may be close to zero; but for other hyperbolas, the angle may be close to 180 degrees, and the hyperbola may be almost a straight line.

ELLIPSE AND CIRCLE

DEFINITION

We begin this section with the definition of an ellipse. An ellipse is the set of all points (x, y) in the plane such that the sum of the distances from (x, y) to two fixed points is some constant. The two fixed points are called the foci, which is the plural of focus.

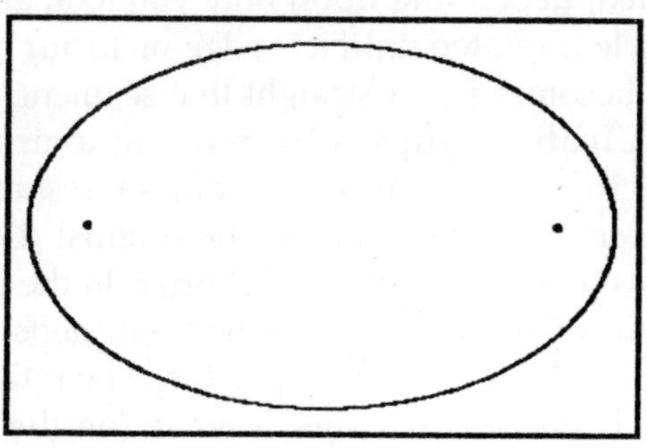

Note that this is clearly a geometrical definition and does not point out the relationship between an ellipse and its algebraic equation.

GENERATING ELLIPSES

Since ellipses can have various shapes, it is important to understand how the various shapes are related to each other, and the terms which are used to descirbe those relationships.

There are a large number of ways in which the generation, or creation, of ellipses can be accomplished. If we want to calculate the exact place where each part of the curve is located, we would use some kind of algebraic curve, such as,

$$(x^2/a^2) + (y^2/b^2) = 1$$

or, as another example, $r = a\ (1 + e \cos\theta\ )$. As useful as these formulae might be for calculations, unless you are very familiar with mathematics, it is difficult to see what the resulting curves look like, or how changing the parameters (the constants) in the equations affects the appearance of the curve. It is easier to understand the nature of ellipses of various types by using graphical methods of creating them, as in the case of cutting a cone with a plane.

One very simple way to generate ellipses is to take a circle and rotate it about a diameter. As shown below, the diameter which is used as an axis of rotation is unchanged by the rotation, but all the diameters which make an angle to that one are reduced in size (foreshortened) by the rotation. If the rotation is small, the resulting ellipse is very nearly round, but if the rotation is large, the ellipse becomes very flattened (or very elongated, depending upon how you look at the effect), and if the circle is rotated until it is edge-on to our line of sight, the "ellipse" becomes just a straight line segment.

Above: Creating ellipses by rotating a circle about a diameter. As the circle rotates, it becomes foreshortened. The axis of rotation stays the same as the original diameter, but the other axes become shorter and shorter. In the process, the ellipse becomes flattened, or stretched out, depending upon your viewpoint. All of the curves produced by rotating a circle are ellipses. Even the extremes, a circle (on the left) and a straight line segment (on the right) are, technically, considered to be ellipses -- just ellipses of rather special type. As you will see, the shape (or, more accurately, the position of the focus) of an ellipse can be expressed by a number, called the eccentricity, which is somewhere between zero and one. If the eccentricity is zero, the ellipse is a circle. If the eccentricity is one, the ellipse is a straight line segment. "Normal" ellipses have values between these extreme values.

Later, we will discuss how we measure the size of an ellipse. As you will see at that time, all the ellipses generated by rotating a circle are the same "size".

Degenerate Ellipses: Sometimes we encounter situations where things become, in a sense, not what we think they should be. The straight line segment which is considered to be an ellipse of eccentricity one is an example of such a situation. Ellipses are curves, and a straight line segment is not a curve, so it is hard to imagine calling it an ellipse of any sort, whatsoever.

However, since we can get such an "ellipse" in the same way as any other ellipse, it seems fair to consider it a special case, in the same way that a circle is a special case; and to designate this, we give it a special name -- a degenerate ellipse. The word degenerate does not, in physical science or mathematics, refer to some kind of moral failing; it means that by pushing something to an extreme limit, it has "degenerated" from something complex, to something much simpler, and at first glance very, very different.

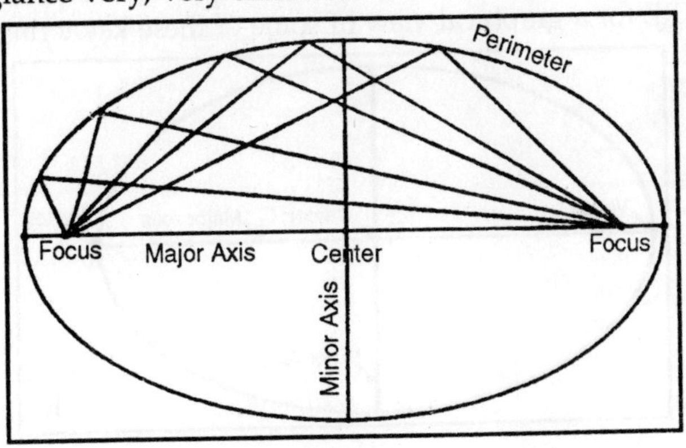

An ellipse can also be created by attaching a string to two thumbtacks, and stretching the string as taut as possible. The resulting curve is the set of all points for which the sum of the distances to the two foci the plural of focus) is a constant. The length of the string is equal to the sum of the distances involved, and is also equal to the major axis of the resulting ellipse (proof to follow). If the two foci are in the same place, the curve that results is a circle, centered on the foci (drawing to follow); if the two foci are so far apart that the string just barely stretches between them, the curve is a straight line

ALGEBRAIC APPROACH

We now move to the algebraic side of ellipses. The standard equation of an ellipse centered at the origin [the point (0,0)] is given by,

$$\frac{x^2}{c^2}+\frac{y^2}{d^2}=1$$

where c and d are some positive constants. The shape of a given ellipse will depend on the relationship between c and d, as will be noted in the examples below. Some other terms need to be introduced at this point of the discussion. The line through the foci intersects the ellipse at two points, known as vertices. This line segment joining the vertices is called the major axis and its midpoint is called the center of the ellipse. The minor axis is the line segment perpendicular to the major axis which also goes through the center and touches the ellipse at two points. See FIGURE for a graphical view of some of these key terms.

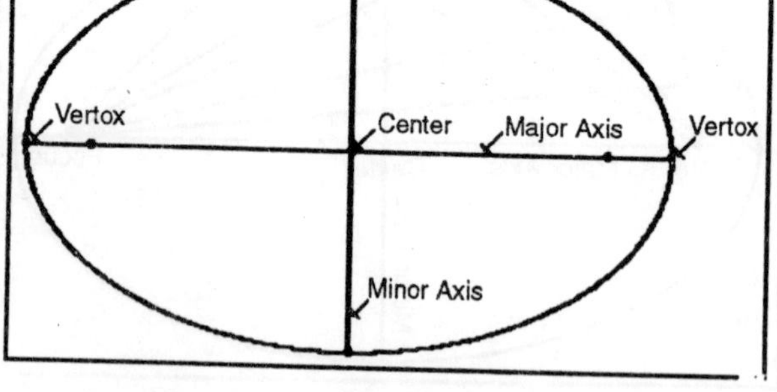

Given these terms, we need to "refine" our discussion of the standard equation of an ellipse. We say that the standard equation of an ellipse centered at the origin is given by,

$$\frac{x^2}{a^2}+\frac{y^2}{b^2}=1$$

if the major axis is horizontal, or,

$$\frac{x^2}{b^2}+\frac{y^2}{a^2}=1$$

if the major axis is vertical. This will insure that we always have the relationship,

$$b<a$$

We can then also note that the distance c of the foci from the center of the ellipse can be determined using a, b, and the following equality:

$$b<a$$

We will use this relationship often, so keep it in mind. The next question you might ask is this: "What happens to the equation if the center of the ellipse is not *(0,0)*?" That is an excellent question. It turns out that the modification in the equation to accomodate this is not difficult to make. If the center of the ellipse is *(n,k)*, then the equation of the ellipse becomes,

$$\frac{(x-h)^2}{a^2}+\frac{(y-k)^2}{b^2}=1$$

if the major axis is horizontal, or,

$$\frac{(x-h)^2}{b^2}+\frac{(y-k)^2}{a^2}=1$$

if the major axis is vertical. One last term related to ellipses should be noted before moving on. This is the concept of the eccentricity of an ellipse. The eccentricity e of an ellipse is defined to be the ratio of the distance between the foci to the distance between the vertices.

Hence,

$$e = \frac{c}{a}.$$

Note that if the foci are close to the vertices, then the eccentricity will be close to 1. (This is visually interpreted as having a fairly flat ellipse.) Note also that e can never be greater than 1 since the foci are always inside the ellipse. This means that the distance between the foci must always be less than the distance between the vertices. Thus, our fraction

$$\frac{c}{a}.$$

will always be less than 1.

What if the foci are very close to one another? Then, c will be quite small and the eccentricity will be very close to 0. Does the eccentricity of an ellipse ever equal 0? The answer is "yes". Remember that a circle is a special case of an ellipse in which, if you will, the two foci have become one. In this case the value of c is 0 (because the distance between the two foci is 0) and this makes,

$$e = 0.$$

Visually, an eccentricity close to 0 then results in a fairly round ellipse. The FIGURE here show three different ellipses with the same vertices but different eccentricities.

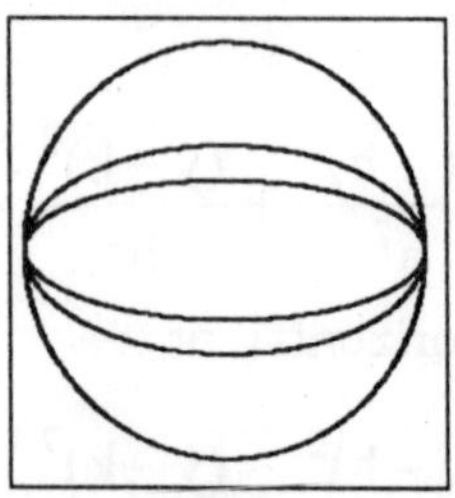

Can you determine which has an eccentricity of 0?

Given all of this theoretical discussion, let's move to some examples.

Example 1: Consider the equation,

$$\frac{x^2}{25} + \frac{y^2}{9} = 1$$

Given our comments above, this equation yields an ellipse.

We see that, $a = 5$ *and* $b = 3$ *and the graph of this ellipse is the following:*

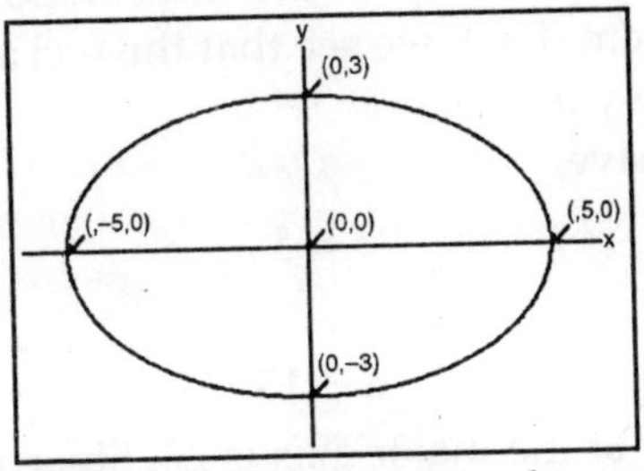

Note the relationship of a and b to the graph. We see that the value of a yields half the length of the major axis. Equivalently, we can say that the vertices are found by traveling down the major axis exactly a units from the center. We also see that half the length of the minor axis is exactly b.

Example 2: Consider the equation:

$$\frac{x^2}{9} + \frac{y^2}{25} = 1$$

It turns out that this ellipse looks very similar to the ellipse in Example 1. The major difference is that this ellipse is now oriented along the vertical axis as opposed to the horizontal. In other words, the major axis of this ellipse is vertical, not horizontal.

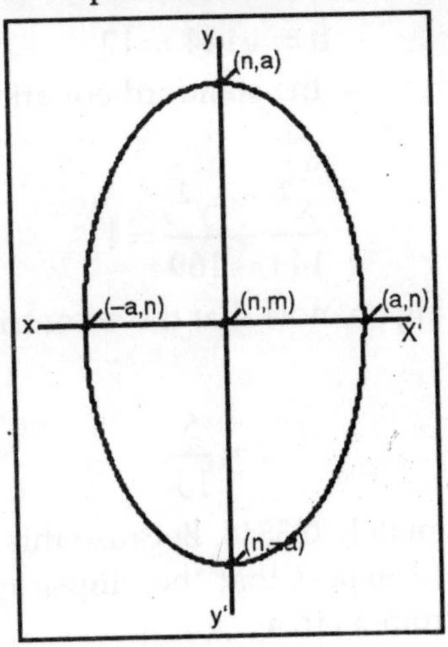

Example 3: Find the standard form of the equation of the ellipse with foci at $(0,5)$ and $(0,-5)$ and with a major axis of length 26. We must interpret the information given to us in the above problem. First, we see that the foci are 10 units apart (and live on the y-axis).

Thus, we have,

$$c = 5\,.$$

Moreover,

$$a = 13$$

since the length of the major axis is 26. Since the center of the ellipse is at (0, 0), the vertices of the ellipse must be at (0, 13) and (0,–13). (Question: How do we know that the vertices are not (13, 0) and(–13, 0)?) Finally, we just need to find out the value of *b*. From the relationship,

$$C^2 = a^2 - b^2,$$

we know that,

$$5^2 = 13^2 - b^2.$$

This can be simplified to,

$$b^2 = 196 - 25 = 144.$$

Thus we see that,

$$b = \sqrt{144} = 12.$$

Now we can write the standard equation of the ellipse. It is,

$$\frac{x^2}{144} + \frac{y^2}{169} = 1\,.$$

Before moving on, note that the eccentricity of this ellipse is,

$$e = \frac{5}{13},$$

which is approximately 0.3846. Because this number is "fairly small", we should expect that the ellipse in question is not much different from a circle.

For a quick look at this ellipse,

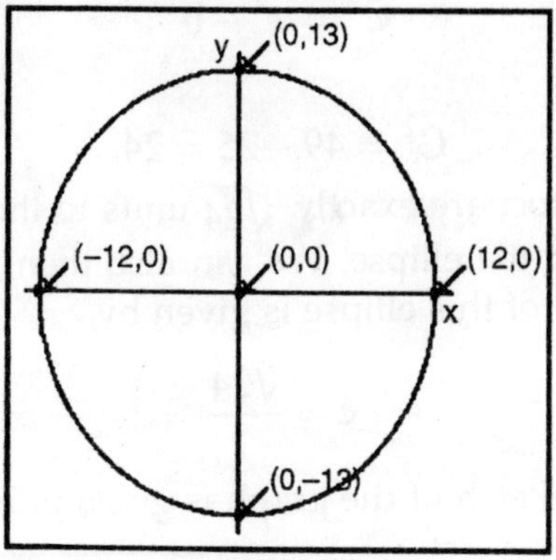

Example 4: Sketch the graph of the ellipse whose equation is,

$$\frac{(x-3)^2}{49}+\frac{(y+1)^2}{25}=1.$$

Again, let's pull as much information out of the equation as possible. We see that the center of the ellipse is (3,–1). (Remember that, $y+1-y-(-1)$ and this is why the y-coordinate of the center is –1.)

Next, note that,

$$a = 7$$

and,

$$b = 5.$$

Since a is in the denominator of the term involving the variable x, we know that the major axis of this ellipse is horizontal (parallel with the x-axis). Moreover, we know that the major axis has length 14 and the vertices occur at points which are 7 units in either direction from the center. This all implies that the vertices are at (3+7,–1) and (3–7,–1), which could also be written as (10,–1) and (–4,–1). As a sidelight, we also know that the endpoints of the minor axis are exactly 5 units above and below the center, which places them at the points (3, 4) and (3, –6). From this information, we can easily plot the ellipse in question. For the sake of completion, let's quickly determine the location of the foci.

Again using the relationship,

$$C^2 = a^2 - b^2$$

we know that,

$$C^2 = 49 - 25 = 24.$$

Thus, the foci are exactly $\sqrt{24}$ units to the left and right of the center of the ellipse. We can also then determine that the eccentricity of this ellipse is given by,

$$e = \frac{\sqrt{24}}{7}$$

Finally, a sketch of the graph is given in Figure.

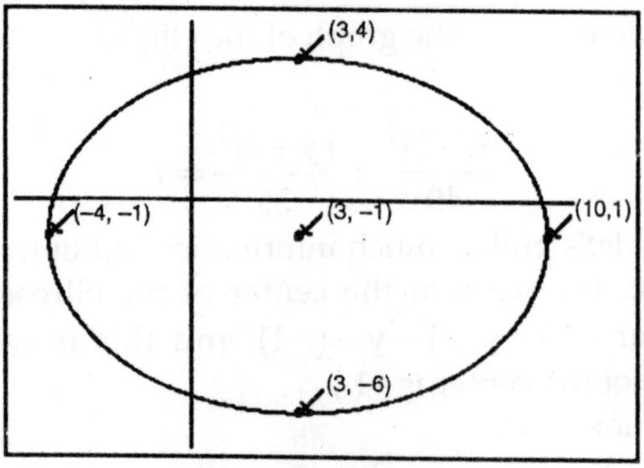

HYPERBOLA

DEFINITION

We begin this section with the definition of a hyperbola. A hyperbola is the set of all points (x, y) in the plane the difference of whose distances from two fixed points is some constant. The two fixed points are called the foci.

Each hyperbola consists of two branches. The line segment which connects the two foci intersects the hyperbola at two points, called the vertices.

The line segment which ends at these vertices is called the transverse axis and the midpoint of this line is called the center of the hyperbola. See Figure H1 for a sketch of a hyperbola with these pieces identified.

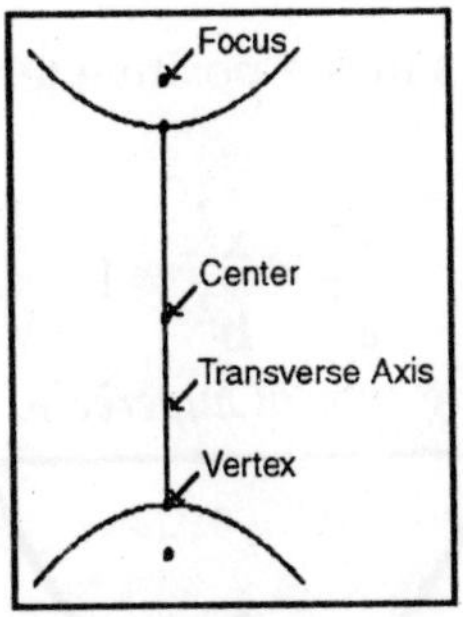

Note that, as in the case of the ellipse, a hyperbola can have a vertical or horizontal orientation. We now turn our attention to the standard equation of a hyperbola. We say that the standard equation of a hyperbola centered at the origin is given by,

$$\frac{x^2}{a^2} - \frac{y^2}{b^2} = 1$$

if the transverse axis is horizontal, or,

$$\frac{y^2}{a^2} - \frac{x^2}{b^2} = 1$$

if the transverse axis is vertical.

Notice a very important difference in the notation of the equation of a hyperbola compared to that of the ellipse. We see that *a always* corresponds to the positive term in the equation of the ellipse. The relationship of *a* and *b* does not determine the orientation of the hyperbola. In the case of the hyperbola, the variable in the "positive" term of the equation determines the orientation of the hyperbola. Hence, if the variable x is in the positive term of the equation, as it is in the equation,

$$\frac{x^2}{a^2} - \frac{y^2}{b^2} = 1,$$

then the hyperbola is oriented as follows:

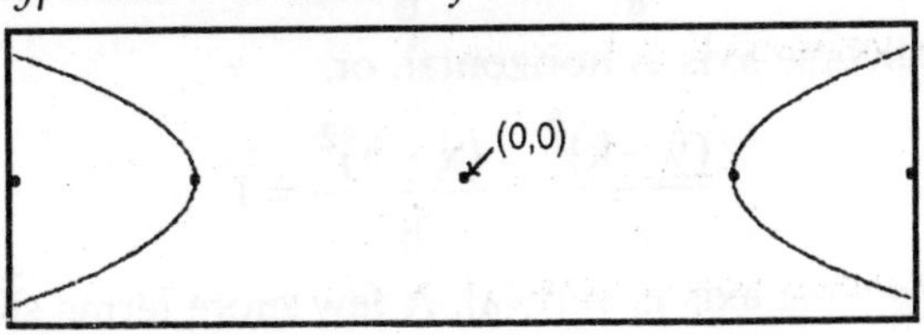

If the variable y is in the positive term of the equation, as it is in the equation,

$$\frac{y^2}{a^2}-\frac{x^2}{b^2}=1,$$

then we see the following type of hyperbola:

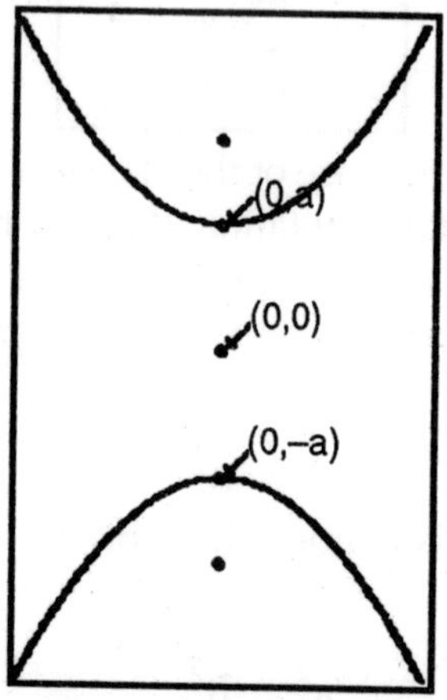

Note that the vertices are always a units from the center of the hyperbola, and the distance c of the foci from the center of the hyperbola can be determined using a, b, and the following equality:

$$b^2=c^2-a^2$$

We will use this relationship often, so keep it in mind. The next question you might ask is this: "What happens to the equation if the center of the hyperbola is not (0,0)?" As in the case of the ellipse, if the center of the hyperbola is (h, k), then the equation of the hyperbola becomes,

$$\frac{(x-h)^2}{a^2}-\frac{(y-k)^2}{b^2}=1$$

if the transverse axis is horizontal, or,

$$\frac{(y-k)^2}{a^2}-\frac{(x-h)^2}{b^2}=1$$

if the transverse axis is vertical. A few more terms should be mentioned here before we move to some examples. First, as in the case of an ellipse, we say that the eccentricity of a hyperbola, denoted by e, is given by,

$$e = \frac{c}{a},$$

or we say that the eccentricity of a hyperbola is given by the ratio of the distance between the foci to the distance between the vertices. Now in the case of a hyperbola, the distance between the foci is greater than the distance between the vertices. Hence, in the case of a hyperbola,

$$e > 1.$$

Recall that for the ellipse,

$$0 \leq e < 1.$$

Two final terms that we must mention are asymptotes and the conjugate axis. The two branches of a hyperbola are "bounded by" two straight lines, known as asymptotes. These asymptotes are easily drawn once one plots the vertices and the points (h, k + b) and (h, k – b) and draws the rectangle which goes through these four points. The line segment joining (h, k + b) and (h, k – b) is called the conjugate axis. The asymptotes then are simply the lines which go through the corners of the rectangle.

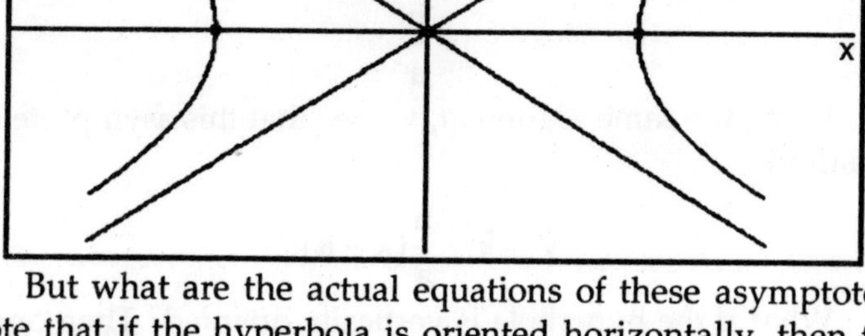

But what are the actual equations of these asymptotes? Note that if the hyperbola is oriented horizontally, then the corners of this rectangle have the following coordinates: (h +

a, k + b), (h – a, k – b), and (h – a, k + b), (h + a, k – b), P> Here I have paired these points in such a way that each asymptote goes through one pair of the points.

Consider the first pair of points:

$$(h + a, k+b), (h - a, k - b),$$

Given two points, we can find the equation of the unique line going through the points using the point-slope form of the line. First, let us determine the slope of our line. We find this as "change in y over change in x" or "rise over run". In this case, we see that this slope is equal to,

$$\frac{2b}{2a}$$

or simply,

$$\frac{b}{a},$$

Then. we also know that the line goes through the center (h, k). Hence, by the point-slope form of a line, we know that the equation of this asymptote is,

$$y - k = \frac{b}{a}(x - h)$$

or

$$y = k + \frac{b}{a}(x - h).$$

The other asymptote in this case has a negative slope which is given by,

$$-\frac{b}{a}.$$

Using the same argument, we see that this asymptote has equation,

$$y = k - \frac{b}{a}(x - h).$$

What if the hyperbola is vertically oriented? Then one of the asymptote will go through the "corners"of the rectangle given by, (h + b, k +a), (h – b, k – a).

Then the slope in this case will not be b/a but will be a/b. Hence, analogous to the work we just performed, we can show that the asymptotes of a vertically oriented hyperbola are determined by,

$$y = k + \frac{a}{b}(x - h)$$

and,

$$y = k - \frac{a}{b}(x - h).$$

Given all of this general information, we move to several examples.

Example 3: Consider the equation,

$$\frac{x^2}{25} - \frac{y^2}{9} = 1.$$

Given our comments above, this equation yields a hyperbola.

We see that,

$$a = 5$$

and

$$b = 3$$

and the graph of this hyperbola is the following:

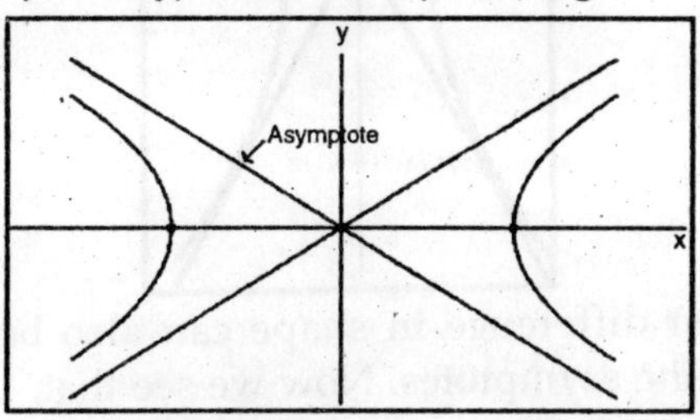

Note also, as we finish this example, that the equations of the asymptotes for this hyperbola are:

$$y = \frac{3}{5}x$$

and

$$y = -\frac{3}{5}x.$$

Example 4. Consider the equation:

$$\frac{x^2}{9} - \frac{y^2}{25} = 1.$$

Note that the only difference between this example and the previous one is that the 9 and 25 have traded places. How does this change the shape of the hyperbola? Is there a change in the orientation from horizontal to vertical? The answer is no. Recall that orientation of a hyperbola is not determined by the sizes of the denominators in the terms of the standard equation of the hyperbola. Rather, orientation is determined by which variable (x or y) is in the ``positive" term. Hence, as is the case in the previous example, this hyperbola is also horizontally oriented. The switch between the 9 and 25 simply changes the shape of the branches.

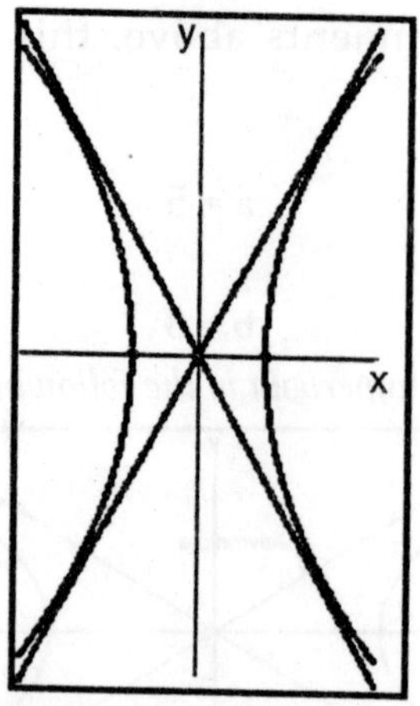

Note that difference in shape can also be seen in the equations of the asymptotes. Now we see that,

$$a = 3$$

and,

$$b = 5.$$

Thus, the slope of the asymptotes is now 5/3, not 3/5 as in the previous example. As the branches of the hyperbolas are drawn and ``squeeze" close to the asymptotes, they will have more room to grow, if you will, than they had in the previous example. This shows us the true effect of switching the places of the 9 and 25 in these two examples.

Example 5: Find the standard form of the equation of the hyperbola with foci at (0, 9) and (0, –9) and transverse axis of length 6. We must interpret the information given to us in the above problem. First, we see that the foci are 18 units apart (and live on the y--axis).

Thus, we have,

$$c = 9.$$

Moreover,

$$a = 3$$

since the length of the transverse axis is 6. Since the center of the hyperbola is at(0, 0), the vertices of the hyperbola must be at (0, 3) and (0, –3). Finally, we just need to find out the value of b. From the relationship,

$$b^2 = c^2 - a^2,$$

we know that,

$$b^2 = 9^2 - 3^2$$

This can be simplified to,

$$b^2 = 81 - 9 = 72.$$

Thus we see that,

$$b = \sqrt{72} = \sqrt[6]{2}.$$

Now we can write the standard equation of the hyperbola. It is,

$$\frac{y^2}{9} - \frac{x^2}{72} = 1.$$

Before moving on, note that the eccentricity of this hyperbola is,

$$e = \frac{9}{3} = 3.$$

Example 6: Determine the center, vertices, foci, and eccentricity of the hyperbola given by the equation

$$\frac{x^2}{4} - \frac{y^2}{49} = 1.$$

Also, determine the equations of the asymptotes of this hyperbola.

Answer: The center of this hyperbola is (0, 0). Since,

$$a = 2$$

and,

$$b = 7,$$

we see that the equation,

$$b^2 = c^2 - a^2,$$

yields,

$$c = \sqrt{53}.$$

Thus, the vertices are located at (–2, 0) and (2, 0), while the foci are located at $(-\sqrt{53}, 0)$ and $(\sqrt{53}, 0)$ (since the hyperbola has a horizontal transverse axis). Also, the eccentricity of this hyperbola is given by,

$$e = \frac{\sqrt{53}}{2}.$$

Finally, the asymptotes are located at,

$$y = \frac{7}{2}x$$

and

$$y = -\frac{7}{2}x.$$

PARABOLA

DEFINITION

A parabola is the set of all points (x, y) that are the same distance from a fixed line (called the directrix) and a fixed point (focus) not on the directrix.

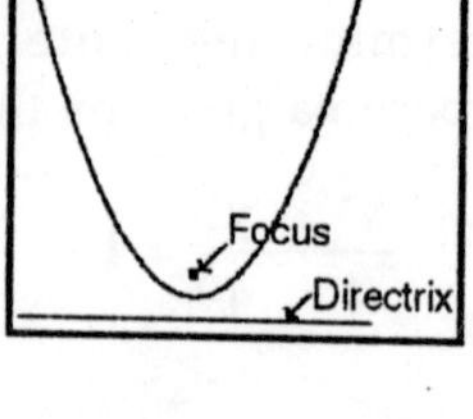

Note that the graph of a parabola is similar to one branch of a hyperbola. However, you should realise that a parabola is not simply one branch of a hyperbola. Indeed, the branches of a hyperbola approach linear asymptotes, while a parabola does not do so. Several other terms exist which are associated with a parabola. The midpoint between the focus and directrix of the parabola is called the vertex and the line passing through the focus and vertex is called the axis of the parabola.

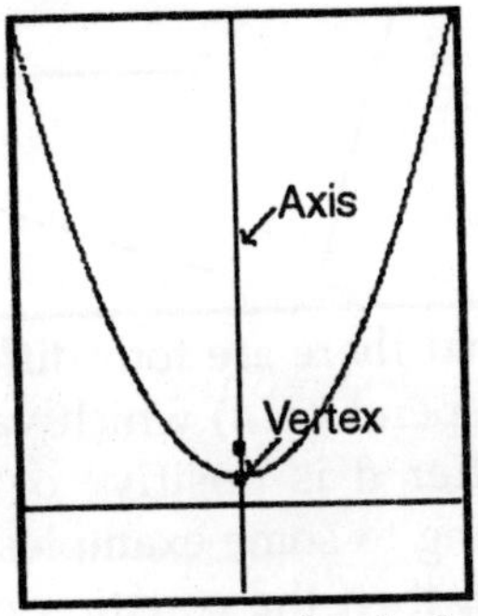

Now let's move to the standard algebraic equations for parabolas and note the four types of parabolas that exist. As we discuss the four types, you should notice the differences in the equations that are related to each of the four parabolas. The standard form of the equation of the parabola with vertex at (0,0) with the focus lying d units from the vertex is given by,

$$x^2 = 4dy$$

if the axis is vertical and,

$$y^2 = 4dx$$

if the axis is horizontal. See Figure for an example with vertical axis for an example with horizontal axis.

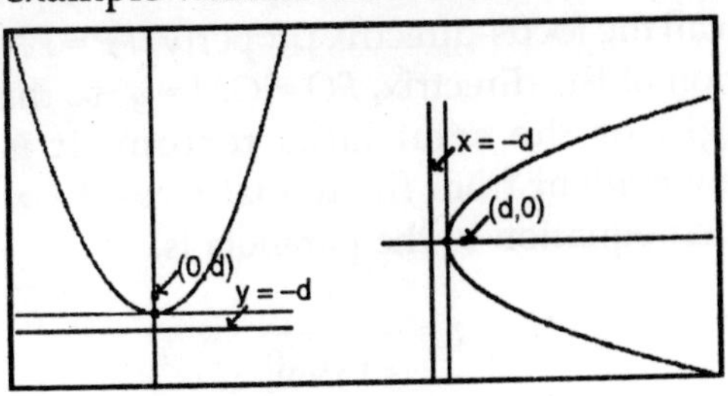

Note here that we have assumed that,

$$d > 0.$$

It is also the case that d could be negative, which flips the orientation of the parabola.

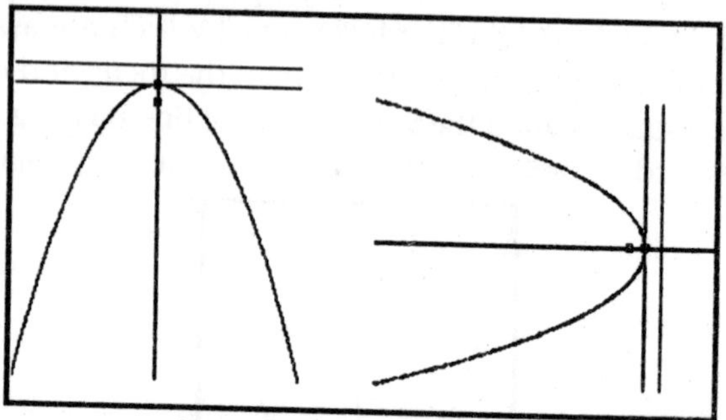

Thus, we see that there are four different orientations of parabolas, which depend on a) which variable is squared (x or y) and b) whether d is positive or negative. One last comment before going to some examples. If the vertex of the parabola is at (h, k), then the equation of the parabola does change slightly. The equation of a parabola with vertex at (h, k) is given by,

$$(x - h)^2 = 4d(y - k)$$

if the axis is vertical and,

$$(y - K)^2 = 4d(x - h)$$

if the axis is horizontal. Now let's move to some specific examples.

PARABOLIC EQUATION

As with the ellipse, we choose the focus as pole and the axis of the parabola as initial line. We shall orient the parabola so that the vertex is towards the right.

We recall the focus-directrix property, $FP = PN$. Also, from the definition of the directrix, $FO = OM = q$, so that $FM = 2q = l$, the length of the semi latus rectum. It is therefore immediately evident from figure that $r \cos \theta + r = 2q = l$, so that the polar equation to the parabola is,

$$r = \frac{l}{1 + \cos\theta}.$$

This is the same as the polar equation to the ellipse, with $e = 1$ for the parabola. I have given different derivations for the ellipse and for the parabola; the reader might like to interchange the two approaches and develop equation in the same manner as we have developed equation.

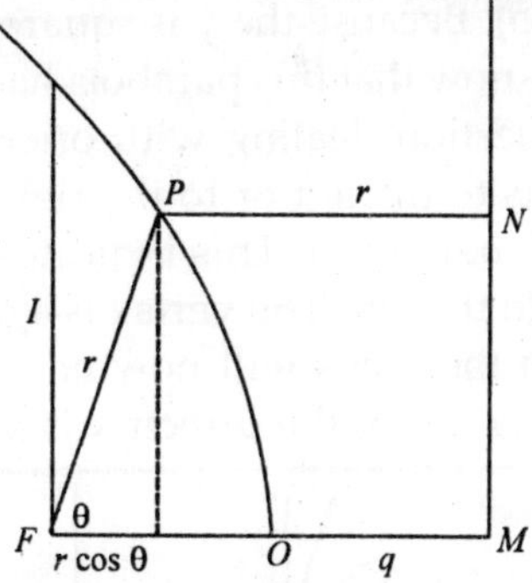

The equation to a *conic section,* and it represents an ellipse, parabola or hyperbola according to whether $e < 1$, $e = 1$ or $e > 1$.

Example 1: *Consider the equation:*

$$x^2 = 8y.$$

Given our comments above, this equation yields a parabola. The value of d in this case is fairly clear; namely,

$$d = 2.$$

The vertex of this parabola is at (0, 0) and the axis of this parabola is vertical (since the variable x is the squared variable.) Thus, the focus of this parabola is situated at the point (0, 2) and the directrix is the horizontal line 2 units below the vertex. Thus, the equation of the directrix is given by $y = -2$.

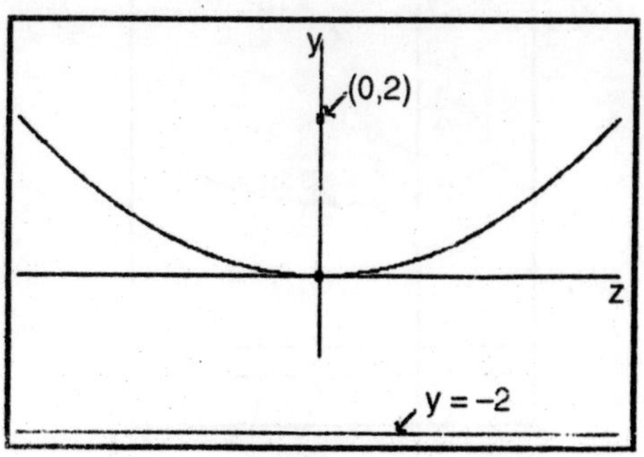

Example 2: Let us now look at the equation of a parabola with horizontal axis. Consider the equation,

$$y^2 = -16x.$$

Again, realise that this is a parabola (as opposed to an ellipse or a hyperbola) because it only has one squared term (as opposed to two). Because the y is squared instead of the x, we automatically know that this parabola has a horizontal axis. The only other question dealing with orientation is whether the parabola opens to the left or to the right.

We note here that d = –4. This indicates that the parabola will indeed open to the left. The vertex is again (0, 0), as in the first example, and the focus will now occur at the point (–4, 0). Finally, the equation of the directrix is x = 4.

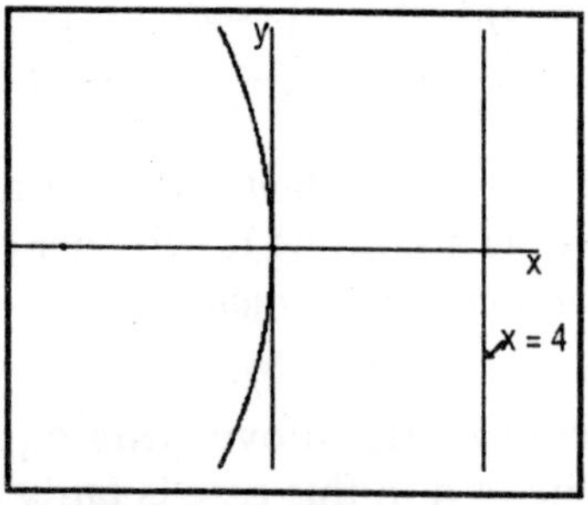

Example 3: *Consider the equation*

$$(x - 3)^2 = 24(y - 5).$$

This is the standard equation of the parabola with vertex (3, 5) and d = 6. Hence, because the axis is vertical, we know that the focus is exactly 6 units above the vertex, placing it at (3, 11). A sketch of this parabola can be found in Figure.

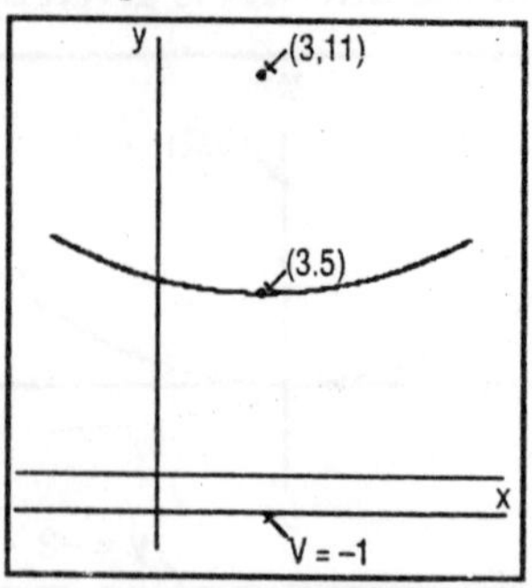

Example 4: Find the standard form of the equation of the parabola with vertex (2, 1) and focus (5, 1).

We see several things from the information given. First, note that the axis of this parabola is oriented horizontally. Next, note that d = 3. Thus, we can quickly write down the standard equation of this parabola. It is,

$$(y - 1)^2 = 12(x - 2).$$

See Figure for a sketch of this parabola.

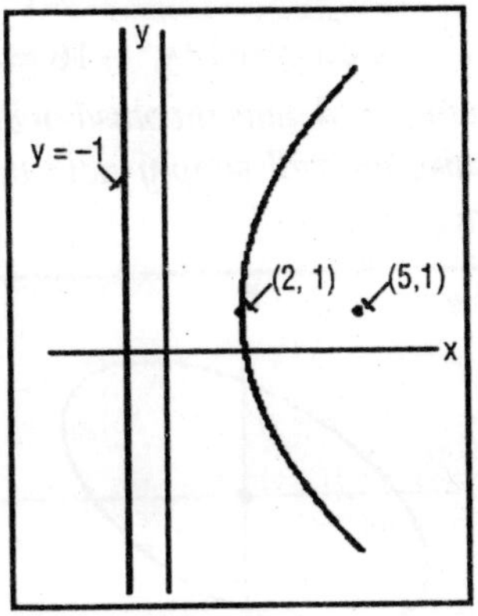

ROTATION OF AXES

Note that in the sections above dealing with the ellipse, hyperbola, and parabola, the algebraic equations that appeared did not contain a term of the form xy. However, in our "Algebraic View of the Conic Sections," we stated that every conic section is of the form,

$$Ax^2 + Bxy + Cy^2 + Dx + Ey + F = 0$$

where A, B, C, D, E, and F are constants. In essence, all of the equations that we have studied have had B=0. So the question arises: "What role, if any, does the xy term play in conic sections? If it were present, how would that change the geometric figure?"

Well, you might also have noticed that for each of the ellipse examples we've considered, the major and minor axes were each parallel to (or coincident with) a coordinate axis; and similarly for the transverse and conjugate axes of each example hyperbola, and the directrix and axis of each parabola.

Is this related to the fact that all of their equations had B=0?

The presence of a nonzero xy term indicates rotation of the graph of the conic section in the plane. For example, in the case of an ellipse, the major axis is no longer parallel to the x-axis or y-axis. Rather, we now have the major axis rotated by an amount that depends on A, B. and C (and which of course is zero when B=0). Let's look at an example.

Example 5: Sketch the graph of the ellipse given by:

$$7x^2 - 6\sqrt{3}xy + 13y^2 - 16 = 0.$$

Without discussing the details involved in determining the angle of rotation in this case, we will simply note that the graph of this ellipse is the following:

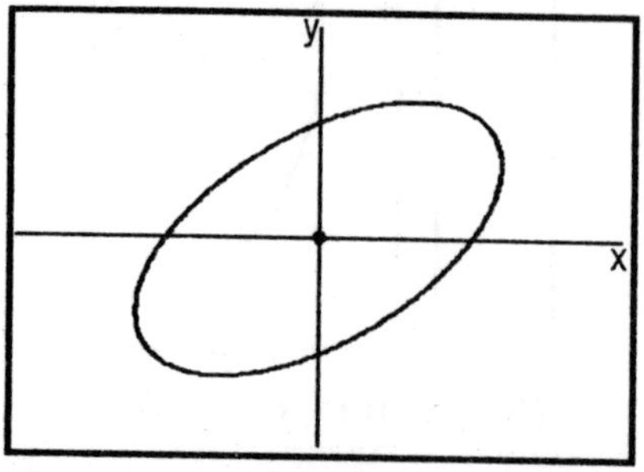

Note that the center of the ellipse happens to be (0,0) and the angle of rotation is,

$$\frac{\pi}{6}$$

In essence, this is simply the graph of the ellipse,

$$\frac{x^2}{4} + \frac{y^2}{1} = 1$$

rotated by an angle of 30 degrees.

CONE

We have so far defined an ellipse, a parabola and a hyperbola without any reference to a cone. However, given the definitions of the ellipse, parabola and hyperbola that we have given, proof is required that they are in fact conic sections. It might also be mentioned at this point that a plane section of a circular cylinder is also an ellipse. Also, of course, if the plane is parallel with the base of the cone, or perpendicular to the axis of the cylinder, the ellipse reduces to a circle.

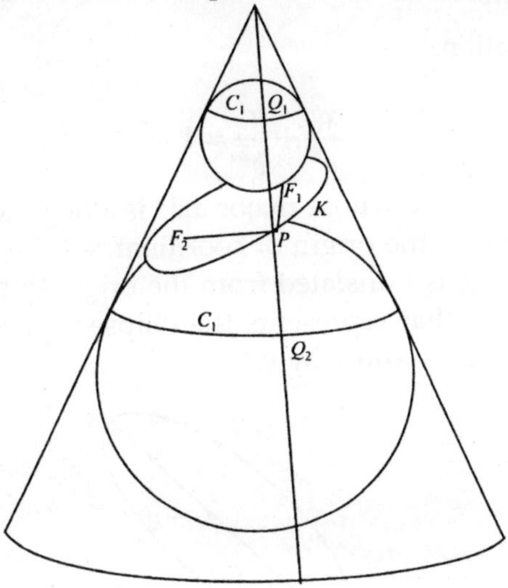

Given: A cone and a plane such that the angle that the plane makes with the base of the cone is less than the angle that the generator of the cone makes with its base, and the plane cuts the cone in a closed curve K.

Required: To prove that K is an ellipse.

Construction: Construct a sphere above the plane, which touches the cone internally in the circle C_1 and the plane at the point F_1. Construct also a sphere below the plane, which touches the cone internally in the circle C2 and the plane at the point F_2. Join a point P on the curve K to F_1 and to F_2.

Draw the generator that passes through the point P and which intersects C_1 at Q_1 and C_2 at Q_2.

Proof: $PF_1 = PQ_1$ (Tangents to a sphere from an external point.) $PF_2 = PQ_2$ (Tangents to a sphere from an external point.)

$$\therefore PF_1 + PF_2 = PQ_1 + PQ_2 = Q_1 Q_2$$

and Q_1Q_2 is independent of the position of P, since it is the distance between the circles C_1 and C_2 measured along a generator.

$\therefore$ K is an ellipse and F_1 and F_2 are its foci.

A similar argument will show that a plane section of a cylinder is also an ellipse.

General Conic

The equation,

$$\frac{x^2}{a^2}+\frac{y^2}{b^2}=1$$

represents an ellipse whose major axis is along the x axis and whose centre is at the origin of coordinates.

If the centre is translated from the origin to the point (p, q), the equation that represents the ellipse will be found by replacing x by $x - p$ and y by $y - q$.

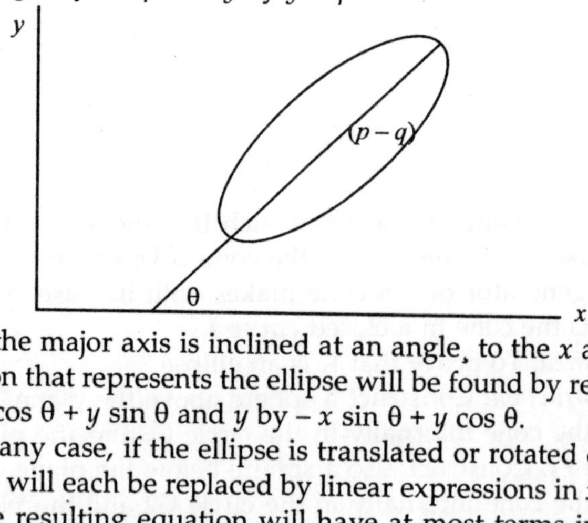

If the major axis is inclined at an angle, to the x axis, the equation that represents the ellipse will be found by replacing x by $x \cos\theta + y \sin\theta$ and y by $-x \sin\theta + y \cos\theta$.

In any case, if the ellipse is translated or rotated or both, x and y will each be replaced by linear expressions in x and y, and the resulting equation will have at most terms in x^2, y^2, xy, x, y and a constant. The same is true of a parabola or a hyperbola. Thus, any of these three curves will be represented by an equation of the form,

$$ax^2 + 2hxy + by^2 + 2gx + 2fy + c = 0.$$

(The coefficients a and b are not the semi major and semi minor axes.) The apparently random notation for the coefficients arises because these figures are plane sections of three-dimensional surfaces (the ellipsoid, paraboloid and hyperboloid) which are described by terms involving the coordinate z as well as x and y.

The customary notation for these three-dimensional surfaces is very systematic, but when the terms in z are removed for the two- dimensional case, the apparently random notation a, b, c, f, g, h remains. In any case, the above equation can be divided through by the constant term without loss of generality, so that the equation to an ellipse, parabola or hyperbola can be written, if preferred, as,

$$ax^2 + 2hxy + by^2 + 2gx + 2fy + 1 = 0.$$

Is the converse true? That is, does an equation of this form always necessarily represent an ellipse, parabola or hyperbola?

Not quite. For example,

$$6x^2 + xy - y^2 - 17x - y + 12 = 0$$

represents two straight lines (it can be factored into two linear terms - try it), while,

$$2x^2 - 4xy + 4y^2 - 4x + 4 = 0$$

is satisfied only by a single point.

However, a plane section of a cone can be two lines or a single point, so perhaps we can now ask whether the general second degree equation must always represent a conic section. The answer is: close, but not quite.

For example,

$$4x^2 + 12xy + 9y^2 + 14x + 21y + 6 = 0$$

represents two parallel straight lines, while,

$$x^2 + y^2 + 3x + 4y + 15 = 0$$

cannot be satisfied by any real (x, y).

However, a plane can intersect a *cylinder* in two parallel straight lines, or a single straight line, or not at all. Therefore, if we stretch the definition of a cone somewhat to include a cylinder as a special limiting case, then we can say that the general second degree equation in x and y does indeed always represent a conic section.

Is there any means by which one can tell by a glance at a particular second degree equation, for example,

$$8x^2 + 10xy - 3y^2 - 2x - 4y - 2 = 0,$$

what type of conic section is represented? The answer is yes, and this one happens to be a hyperbola. The discrimination is made by examining the properties of the determinant,

$$\Delta = \begin{vmatrix} a & h & g \\ h & b & f \\ g & f & c \end{vmatrix}$$

A table after the design of the dichotomous tables commonly used by taxonomists in biology, in which the user is confronted by a couplet (or sometimes triplet) of alternatives, and is then directed to the next part of the table. The symbol $\bar{a}$ to mean the cofactor of a in the determinant, $\bar{h}$ the cofactor of h, $\bar{g}$ the cofactor of g, etc. Explicitly, these are,

$$\begin{aligned} \bar{a} &= bc - f^2, \\ \bar{b} &= ca - g^2, \\ \bar{c} &= ab - h^2, \\ \bar{f} &= gh - af, \\ \bar{g} &= hf - bg \end{aligned}$$

and
$$\bar{h} = fg - ch.$$

The first column labels the choices that the user is asked to make. At the beginning, there are two choices to make, 1 and 1' The second column says what these choices are, and the fourth column says where to go next. Thus, if the determinant is zero, go to 2; otherwise, go to 5. If there is an asterisk in column 4, you are finished. Column 3 says what sort of a conic section you have arrived at, and column 5 gives an example.

No matter what type the conic section is, the coordinates of its centre are $(\bar{g}/\bar{c}, \bar{f}/\bar{c})$ and the angle θ that its major or transverse axis makes with the x axis is given by,

$$\tan 2\theta = \frac{2h}{a-b}.$$

Thus if x is first replaced with $x + \bar{g}/\bar{c}$ and y with $y + \bar{f}/\bar{c}$, and then the new x is replaced with $x \cos\theta - y \sin\theta$ and the new y with $x \sin\theta + y \cos\theta$, the equation will take the familiar form of a conic section with its major or transverse axis coincident with the x axis and its centre at the origin.

FIVE POINTS CONICS

Problem: Draw a conic section through the five points. The first thing to notice is that, since a conic section is of the form

$$ax^2 + 2hxy + by^2 + 2gx + 2fy + 1 = 0$$

(remember that there is no loss of generality by taking the constant term to be 1), five points are necessary and sufficient to define a conic section uniquely. One and only one conic section can be drawn through these five points. We merely have to determine the five coefficients.

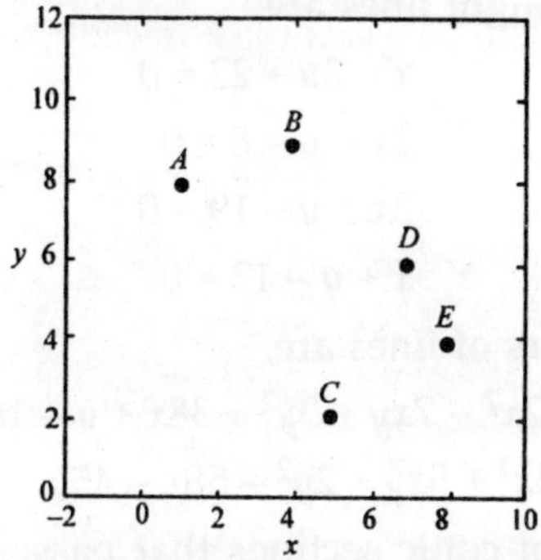

Fig. The Five Points $A(1, 8)$, $B(4, 9)$, $C(5, 2)$, $D(7, 6)$, $E(8, 4)$.

The most direct (but not the fastest or most efficient) way to do this is to substitute each of the (x, y) pairs into the equation in turn, thus obtaining five linear equations in the five coefficients. There is a better way. We write down the equations for the straight lines AB, CD, AC and BD. Let us call these equations $\alpha = 0$, $\beta = 0$, $\gamma = 0$ and $\delta = 0$ respectively.

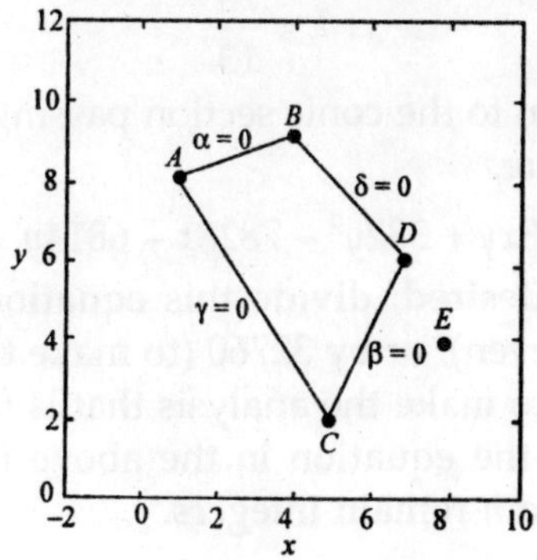

Then $\alpha\beta = 0$ is the equation that represents the two straight lines AB and CD, and $\gamma\delta = 0$ is the equation that represents the two straight lines AC and BD. The equation $\alpha\beta + \lambda\gamma\delta = 0$, where λ is an arbitrary constant, is a second degree equation that represents any conic section that passes through the points A, B, C and D. By inserting the coordinates of E in this equation, we can find the value of that forces the equation to go through all five points. This model of unclarity will become clear on following an actual calculation for the five points of the present example.

The four straight lines are,

$$\alpha = 0: \qquad x - 3y + 23 = 0$$

$$\beta = 0: \qquad 2x - y - 8 = 0$$

$$\alpha = 0: \qquad 3x + y - 19 = 0$$

$$\alpha = 0: \qquad x + y - 13 = 0$$

The two pairs of lines are,

$$\alpha\beta = 0: 2x^2 - 7xy + 3y^2 + 38x + y - 184 = 0$$

$$\gamma\delta = 0: 3x^2 + 5xy + 2y^2 - 58x - 45y + 247 = 0$$

and the family of conic sections that passes through A, B, C and D is,

$$\alpha\beta + \lambda\gamma\delta = 0$$

$$(2 + 3\lambda)x^2 - (7 - 5\lambda)xy + (3 + 2\lambda)y^2 + (38 - 58\lambda) + (1 - 45\lambda)y - 184 + 247\lambda = 0.$$

Now substitute $x = 8$, $y = 4$ to force the conic section to pass through the point E. This results in the value,

$$\lambda = \frac{76}{13}.$$

The equation to the conic section passing through all five points is therefore,

$$508x^2 + 578xy + 382y^2 - 7828x - 6814y + 32760 = 0$$

We can, if desired, divide this equation by 2 (since all coefficients are even), or by 32760 (to make the constant term equal to 1) but, to make the analysis that is to follow easier, I choose to leave the equation in the above form, so that the constants f, g and h remain integers.

The constants have the values, $a = 508$, $b = 382$, $c = 32760$, $f = -3407$, $g = -3914$, $h = 289$ and the cofactors have the values,

$$\bar{a} = 906\,671, \bar{b} = 1\,322\,684, \bar{c} = 110\,535$$

$$\bar{f} = 599\,610, \bar{g} = 510\,525, \bar{h} = 867\,358$$

Let us consult the dichotomous table. The value of the determinant is $\Delta = a\bar{a} + b\bar{b} + c\bar{c}$ (or $h\bar{h} + b\bar{b} + f\bar{f}$, or $g\bar{g} + f\bar{f} + c\bar{c}$; try all three sums to check for arithmetic mistakes). It comes to $\Delta = -419\,939\,520$, so we proceed to option 5. $\bar{c} > 0$,so we proceed to option 6. a and Δ have opposite signs, so we proceed to 7. a does not equal b, nor is h equal to zero. Therefore we have an ellipse.

The centre of the ellipse is at (4.619, 5.425), and its major axis is inclined at an angle 128° 51' to the x-axis. If we now substitute $x + 4.619$ for x and $y + 5.425$ for y, and then substitute x cos 128°51' + y sin 1280 51' for the new value of x and $-x$ sin 128° 51' + y cos 128° 51' for the new value of y, the equation will assume its the familiar form for an ellipse referred to its axes as coordinate axes and its centre as origin.

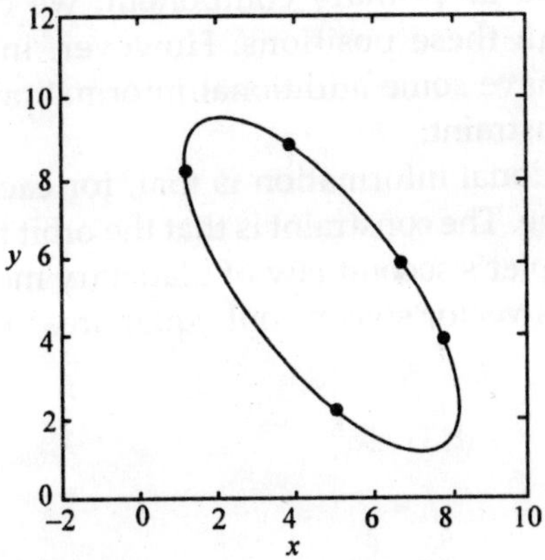

Tracing of Conic Section through Points

What is the best ellipse passing near to the following 16 points?

(1, 50) (11, 58) (20,63) (30,60)

(42, 59) (48, 52) (54, 46) (61, 42)

(61, 19) (45, 12) (35, 10) (25, 13)

(17, 17) (14, 22) (5, 29) (3, 43)

This is answered by substituting each point (x, y) in turn in the equation,

$$ax^2 + 2hxy + by^2 + 2gx + 2fy + 1 = 0,$$

thus obtaining 16 equations in the coefficients a, h, b, g, f. (The constant term can be taken to be unity.) These are the equations of condition. The five normal equations can then be set up and solved to give those values for the coefficients that will result in the sum of the squares of the residuals being least, and it is in that sense that the "best" ellipse results. The details of the method are given in the chapter on numerical methods. The actual solution for the points given above is left as an exercise for the energetic.

It might be thought that we are now well on the way to doing some real orbital theory. After all, suppose that we have several positions of a planet in orbit around the Sun, or several positions of the secondary component of a visual binary star with respect to its primary component; we can now fit an ellipse through these positions. However, in a real orbital situation we have some additional information as well as an additional constraint.

The additional information is that, for each position, we also have a time. The constraint is that the orbit that we deduce must obey Kepler's second law of planetary motion - namely, that the radius vector sweeps out equal areas in equal times.

3

Differential Geometrical Sphere

INTRODUCTION

We are interested here in the geometry of an ordinary sphere. In plane geometry we study points, lines, triangles, polygons, etc. On the sphere we have points, but there are no straight lines – at least not in the usual sense. However, straight lines in the plane are characterized by the fact that they are the shortest paths between points. The curves on the sphere with the same property are the great circles. Therefore it is natural to use great circles as replacements for lines. Then we can talk about triangles and polygons and other geometrical objects. In these notes we will do this, and at the same time we will continuously look back to the plane to compare the spherical results with the planar results.

We will study the incidence relations between great circles, the notion of angle on the sphere, and the areas of certain fundamental regions on the sphere, culminating with the area of spherical triangles. Our ultimate goal is two very nice results. First we will prove Girard's Theorem, which gives a formula for the sum of the angles in a spherical triangle. Then we will use Girard's Theorem to prove Euler's Theorem that says that in any convex, three dimensional polyhedron we have $V - E + F = 2$, where V is the number of vertices, E is the number of edges, and F is the number of faces.

INFORMATION

We will start with the basics. A sphere is a set of points in three dimensional space equidistant from a point called the center of the sphere. The distance from the center to the points on the sphere is called the radius of the sphere. Notice that we are talking about the surface of a ball, and not the ball itself. The surface of the earth we live on is a good approximation to a sphere. As you read through this material it would be helpful to have a beach ball about 12" in diameter with a smooth, solid coloured surface, a marking pen, preferably the kind that easily washes off, string, scissors, and a protractor.

A sphere of radius R, a triangle Δ ABC with interior angles α, β, and γ has area given by: arca ($\Delta\, ABC$) = R^2 ($\alpha + \beta + \gamma - \pi$)- First, some terminology: a lune is one of the regions between two great circles and the antipodal points where they cross. Antipodal points or figures are those that correspond to opposite ends of a diameter of the sphere. A lune is given by the angle q at the centre of the sphere between the two defining planes.

Proof: From our assumptions above, Arca of the lunc,

$$= \frac{\theta}{2\pi} \cdot 4\pi R^2 = 2\theta R^2$$

To make our proof easier, we may assume that the triangle lies in one hemisphere. If not you can use Assumption to break the triangle up into smaller triangles that do lie in one hemisphere. A triangle is the intersection of three lunes: the lune given by $\overrightarrow{AB} \cap \overrightarrow{AC}$ the lune given by $\overleftrightarrow{AB} \cap \overleftrightarrow{AC}$ and the lune given by Each of these three lunes has an antipodal lune and the intersection of these three antipodal lunes is an antipodal triangle Δ *A¢ B¢ C¢*.

We will denote the lunes by the angle that includes them, *BCA*, *BAC*, and *ABC*. Starting with the triangle and the lune *ABC*, add to them the lune *ACB*. This covers the triangle Δ *A′ B′ C′* three times. Now, add in the antipodal lune to lune *CAB* and the antipodal triangle Δ *A′ B′ C′*. This covers the hemisphere determined by the line $\overleftrightarrow{BC}$ counting the triangle three times.

So we get:

$$2\pi R^2 = 2\alpha R^2 + 2\gamma R^2 - 3.\ \text{arca}\ (\Delta ABC) + \text{arca}\ (\Delta ABC)$$

Thus, we get our desired formula: Arca (ΔABC) = R^2 ($\alpha + \beta + \gamma - \pi$). We call the value of $\alpha + \beta + \gamma - \pi$ the angle excess of the spherical triangle. Note that since every triangle on the sphere has area, every triangle on the sphere has interior angle sum greater than If the radius of the sphere is very large and the triangle very small in area, then the angle excess will be almost 0. Let Σ_R denote the sphere of radius R centered at the origin in R^3. In rectangular coordinates,

$$\Sigma_R = \{(x, y, z) \mid x^2 + y^2 + z^2 = R^2\}$$

. *And in spherical coordinates*: $\Sigma R = \{(R \cos(\theta) \sin(\phi), R \sin(\theta) \sin(\phi), R \cos(\phi) \mid 0 \le \theta \le 2\pi, 0 \le \phi \le \pi\}$.

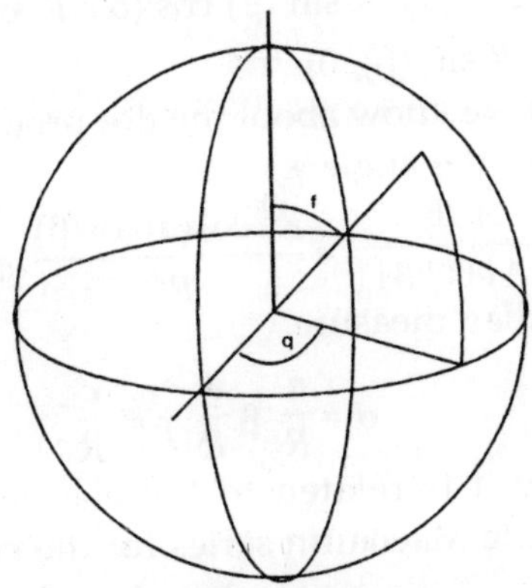

Fig. Spherical Coordinates

The lengthof a path between two points along a great circle is found by measuring the angle, in radians, made by the radii at the two endpoints and multiply it by the radius of the sphere, *i.e.* $R\theta$. Note that the length of a segment is unchanged when we rotate the sphere around some axis or reflect it through a great circle.

PYTHAGOREAN THEOREM

For a right triangle ΔABC on a sphere of radius R with a right angle at vertex C and sides of length a, b, and c, then,

$$\cos\left(\frac{C}{R}\right) = \cos\left(\frac{a}{R}\right) \cos\left(\frac{b}{R}\right)$$

Proof: Rotate the sphere so that the point A has coordinates $(R,0,0)$ and the point C lies in the xy-plane. This will make β

and $\frac{\pi}{2}$ the angles determining the point B. Here β is the central angle determined by side AC, α is the angle determined by AC and γ is the angle determined by AB. This gives us the following spherical coordinates for the vertices of the triangle: $A = (R, 0, 0)$.

$$B = \left(R \cos(\beta) \sin(\frac{\pi}{2} - \alpha), R \sin(\beta) \sin(\frac{o}{2} - \alpha), R \cos(\frac{\pi}{2} - \alpha)\right)$$

$$= (R \cos(\beta) \cos(\alpha), R \sin(\beta) \cos(\alpha), R \sin(\alpha))$$

$$c = (R \cos(\beta), R \sin(\beta), 0)$$

Using what we know about the dot product in R^3, we can find the cosine of the angle γ.

$$\cos(\gamma) = \frac{A.B}{||A||||B||} = \frac{R^2 \cos(a)\cos(\beta)}{R^2} = \cos(\alpha)\cos(\beta).$$

Now, in radian measure,

$$\alpha = \frac{a}{R} \; \beta \frac{b}{R} \; \gamma = \frac{c}{R}$$

To see how it is related to the classical Pythagorean theorem, recall the Maclaurin series for the cosine function:

$$\cos(x) = 1 - \frac{x^2}{2!} + \frac{x^4}{4!} - \frac{x^6}{6!} + \ldots$$

Using this in the formula from the spherical Pythagorean theorem gives us:

$$1 - \frac{c^2}{2R^2} + \frac{c^4}{24R^4} - \ldots = \left(1 - \frac{a^2}{2R^2} + \frac{a^4}{24R^4} - \ldots\right)\left(1 - \frac{b^2}{2R^2} + \frac{b^4}{24R^4} - \ldots\right)$$

$$= 1 - \frac{a^2}{2R^2} - \frac{b^2}{2R^2} + \frac{a^2b^2}{4R^4} + \ldots$$

$$c2 + \frac{\text{tcrms in higher powers of } c}{R^2} = a2 + b2 + \frac{\text{tcrms in higher powers of } a \text{ and } b}{R^2}$$

Now, as R gets arbitrarily large while a, b, and c stay small, we get the classical Pythagorean theorem as a limiting case.

LINES AND SPHERES

If we take an arbitrary line and a sphere in three space, several things can happen. First, the line and the sphere can miss each other. That case is not very interesting. Secondly, the line can intersect the sphere in only one point. In that case the line is tangent to the sphere. The only other thing that can happen is that the line hits the sphere in precisely two points. In particular a line cannot lie in the sphere.

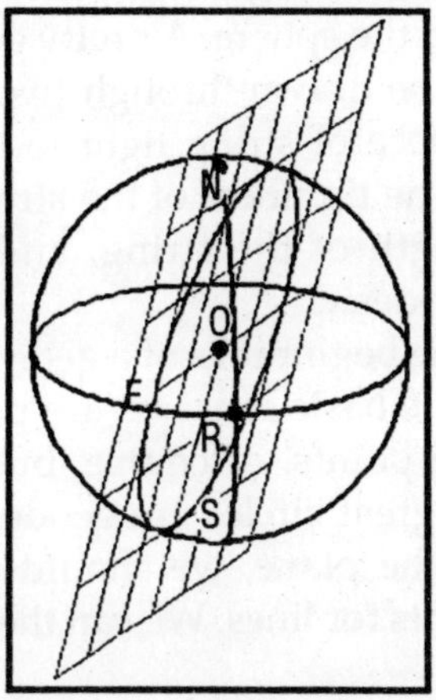

The case which is most interesting is when the line passes through the center of the sphere. In this case the two points of intersection with the sphere are said to be antipodal points. The best known example of antipodal points is the north and south poles on the earth.

PLANES, SPHERES AND CIRCLES

Next let's look at a plane and a sphere. Again there are several things that can happen. In the uninteresting case the plane and the sphere miss each other. If they do meet each other there are two possibilities. First they can meet in a single point. In this case the plane is tangent to the sphere at the point of intersection. In the other case the sphere and the plane meet in a circle.

It is easy to see that the circle of intersection will be largest when the plane passes through the center of the sphere, as it does in the figure to the left. Such a circle is called a great circle. A geographic example of a great circle is the equator. The meridians of longitude form exactly half a great circle. The parallels of latitude are small circles, except for the equator.

Great circles become more important when we realise that the shortest distance between two points on the sphere is along the segment of the great circle joining them. On any surface the curves that minimize the distance between points are called geodesics. Thus lines are the geodesics on the plane, and great circles fill that role on the sphere. A pretty good approximation to a great circle can be drawn through two points on a beach ball by holding a piece of string tight to the ball at the two points in question. The tightness of the string has the effect of minimizing the length of the string, and therefore closely approximating a geodesic.

We now have the beginnings of a geometry on the sphere. In plane geometry the basic concepts are points and lines. On the sphere we have points, of course, but no lines as such. However, since the great circles are geodesics on the sphere, just as lines are in the plane, we should consider the great circles as replacements for lines. We can then compare the two geometries.

Exercise: Take your favourite geometry book, and find its list of axioms. Which of the axioms are true for the sphere? If a planar axiom is not literally true on the sphere try to rephrase it so it is true.

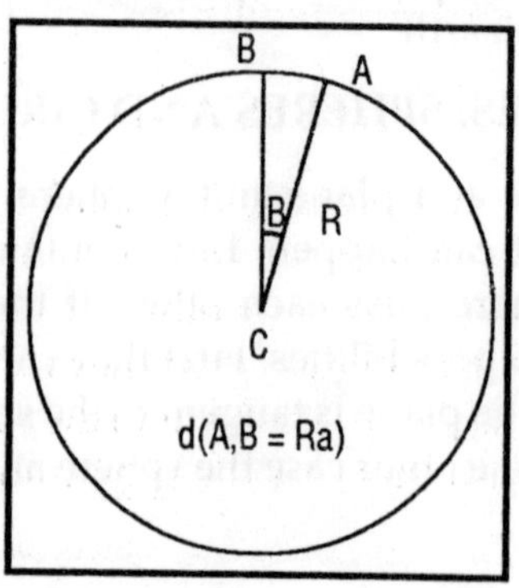

This exercise cannot be completed at this point. It is one that will continue throughout this material. At this point you should check the axioms of incidence. Remember that a great circle is the intersection of the sphere with a plane that passes through the center of the sphere. Almost every fact we will discover about great circles will follow from that fact.

INCIDENCE RELATIONS ON A SPHERE

Suppose we have two distinct points A and B on the sphere. Together with C, the center of the sphere, we have three points in space, and there are two possibilities. First suppose that A and B are not antipodal points. Then A, B, and C do not lie on the same line in space, and consequently determine a unique plane.

This plane passes through C, the center of the sphere, and consequently the intersection of the plane with the sphere is a great circle containing A and B. Thus A and B determine a unique great circle.

If A and B are antipodal points, then A, B, and C lie on the same line in space. Any plane which contains this line determines a great circle which must contain A and B. Thus there are infinitely many great circles containing A and B if they are antipodal.

To sum up, the first incidence relation for the sphere is:

- If A and B are two points which are not antipodal, then there is a unique great circle that contains both of them. If A and B are antipodal, then there are infinitely many great circles conatining them.

Now suppose we have two great circles. Each of these is the intersection of the sphere with a plane through the center. These planes must intersect in a line in space, which of course interesects the sphere in two antipodal points.

Thus the second incidence relation is:

- Two distinct great circles meet in exactly two antipodal points.

Here is an easy question that you should be able answer. Are there parallel great circles on a sphere? To answer you will need to decide what parallel means on the sphere.

SPHERICAL DISTANCE AND ISOMETRIES

If A and B are two points on the sphere, then the distance between them is the distance along the great circle connecting them. Since this circle lies totally in a plane, we can figure this distance using the plane figure to our left. If the angle ACB is a, and if a is measured in radians, then the distance between A and B is given by,

$$d(A,B) = R\,a,$$

where R is the radius of the sphere. An isometry of the sphere is a mapping of the sphere to itself which preserves the distance between points. It is easy to see that a rotation of the sphere around one of the sphere's diameters is an isometry. It simply rotates the picture to the left into another one just like it, but in a different plane. Another example of an isometry is the antipodal map, which maps a point onto the point on the other side of the sphere. In other words, given a point A on the sphere, its image under the antipodal map is the other intersection of the line AC through the point A and the center of the sphere C, with the sphere.

LUNES

In the plane the simplest polygon is the triangle. There are no interesting polygons with only two sides. This is not true on the sphere. Any two great circles meet in two antipodal points, and divide the sphere into four regions each of which has two sides which are segments of great circles. We will call such a region a lune, or a biangle.

Why is it called a lune? The name comes from the Latin word luna, which means moon. Think about the part of the moon that is seen at any time. That portion has to be in the hemisphere which is illuminated by the sun, and in the hemisphere that is visible from the earth. The intersection of two hemispheres is precisely a lune. Lunes are pretty simple things. However there are two things we should notice about them.

- The vertices of a lune are antipodal points.
- The two angles of a lune are equal.

ANGLES ON THE SPHERE

What do we mean by an angle on the sphere? How do we measure them? After all curves on a sphere do not lie in a plane. However, the lines that are tangent to the two intersecting curves are both in the plane that is tangent to the sphere at the point of intersection. We define the angle between two curves to be the angle between the tangent lines. We should mention that in these notes all angles will be measured in radians.

With a protractor and a little practise it is possible to measure spherical angles pretty accurately. In the case of a lune, the angle between the great circles at either of the vertices is simply the angle between the planes that define the great circles, and so it does not matter at which vertex the measurement is made.

AREA ON THE SPHERE

The area of a sphere of radius R is $4\pi R^2$. A great circle divides the sphere into two congruent hemispheres. Each of these will have area $4\pi R^2$. Another great circle, which meets the first at right angles, divides the sphere into four congruent lunes, each with area $4\pi R^2$.

We can continue this process by dividing each of these four lunes into two by bisecting the angle. We then get eight congruent lunes, each with area $\pi R^2/2$. Notice that the lunar angle for each of these lunes is $\pi/4$ radians, or 45 degrees.

AREA OF A LUNE

Suppose we divide a hemisphere into q equal lunes by drawing great cirles all from one point on the great circle which froms the boundary of the hemisphere. The lunar angle of each is π/q, and the area of each lune is $2\pi R^2/q$. Now if we look at the union of p of these lunes, we get one lune with lunar angle $p\pi/q$, and the area will be $2p\pi R^2/q$.

Thus if $\alpha = p\pi/q$ is the lunar angle, then the area of the lune is $2R^2\alpha$.

Thus,

$$\text{area(lune)} = 2R^2(\text{lunar angle}).$$

We have proved this for angles of the form $\alpha = p\pi/q$, but from this it follows for general angles by approximating the angle with rational multiples of π.

It is interesting to consider the formula for the area of a lune when the angle is being measured in degrees instead of radians. Since one radian is equal to $180/\pi$ degrees, and one degree is equal to $\pi/180$ radians, the formula becomes $\text{area}(L_\alpha) = 2R^2 \times \alpha \times \pi/180 = \pi R^2/90 \times \alpha$. Comparison of the formulas provides yet another reason for using radians.

SPHERICAL TRIANGLES

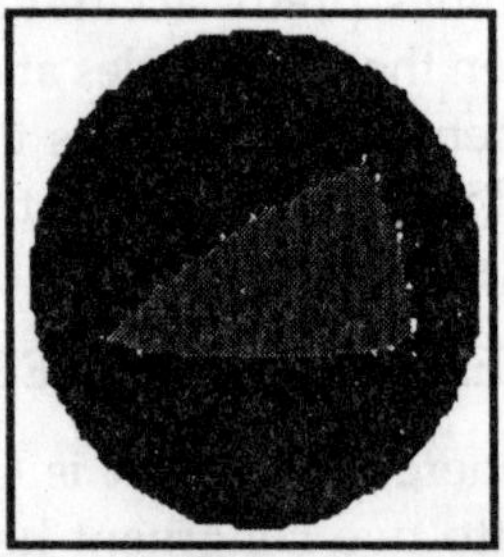

A spherical triangle is defined just like a planar triangle. It consists of three points called vertices, the arcs of great circles that join the vertices, called the sides, and the area that is inclosed therein. However as soon as that is said you realise that there is a lot of ambiguity. First of all, there are two segments of the great circle that join each pair of points. Then, once we have decided which of the two segments to use, the resulting figure is the boundary of two different regions --- what might be called the outside as well as the inside. Since only one "long" side can be used without getting extra points of intersection of the sides, we can count 8 regions that might be called triangles, all of which have these three points as vertices. Of course the situation is even more complicated if two of the points are antipodal.

To get away from this complexity, we will deal only with small triangles. Given the three vertices, no pair of which are antipodal, the small triangle has as sides the short segments of great circles that join the vertices. We notice that there is a hemisphere that contains the three points, and the region we choose is the region which is bounded by the sides and lies entirely in that hemisphere. Actually the results we will discover are true for "large" triangles, but we will leave the verification of that to the reader.

Exercise: Experimenting with Spherical Triangles

Draw three spherical triangles of different sizes on a beach ball (Boyd Hemphill has a way to increase the precision with which this can be done). Measure the angles of the triangles and compare the sum of the angles with the relative size of the triangle. If there are others working on this, compare your results with theirs. Do you notice any pattern in the data? In particular how does the sum compare with the 180 degrees we expect in the plane? What happens to the sum of the angles as the size of the triangle gets larger?

Exercise: The area of a Spherical Triangle

Can you find a formula that relates the area of a spherical triangle to the sum of its angles? This is a difficult problem, but you know everything you need to know to solve it. Think about the three sides of the triangle and the great circles of which they are a part. For each angle of the triangle, these great circles bound two congruent lunes, one of which contains the triangle and one which does not. If you can visualize the configuration of these six lunes on the sphere, you will be on the way to a solution.

SPHERE: CONSEQUENCES OF GIRARD'S THEOREM

Exercise: Distortion in Maps

Everyone knows that a map from even a small portion of the sphere to the plane must involve some distortion. However Girard's Theorem makes that statement more precise. Let's call a map ideal if does two things.

- It maps great circles to straight lines.
- It preserves angles.

Exercise: Similarity

Suppose we have a triangle on the sphere with angles A,

B, and C. Can we find a larger triangle with the same angles? In other words, do similar triangles exist on the sphere? The answer is no! By the area formula, any triangle with these angles must have the same area, and therefore cannot be larger.

Exercise: Congruence Theorems

In the plane there are a number of theorems aobut the congruence of triangles. They are usualy referred to by their acronyms, i.e., SSS, SAS, AAS, and ASA. Can you modify the proofs of the planar theorems to prove these theorems on the sphere? It might be necessary to modify the statements of the theorems to takeical Sine Theorem.

Theorem: Let $\Delta\, ABC$ be a spherical triangle on a sphere of radius R. Let a, b, and c denote the lengths of the sides in radians, and let $\angle A$, $\angle B$, and $\angle C$ denote the interior angles at each vertex.

Then

$$\frac{\sin\left(\frac{a}{R}\right)}{\sin(\angle A)} = \frac{\sin\left(\frac{b}{R}\right)}{\sin(\angle B)} = \frac{\sin\left(\frac{c}{R}\right)}{\sin(\angle C)}$$

Proof: We will first prove this for a right triangle and then use that result to prove the general case. In addition we can restrict our attention to triangles that lie entirely within a quarter of a hemisphere.

Let $\Delta\, ABC$ be a right triangle with right angle at C. Extend the radius OA to OA', where A' is the on the intersection of the tangent plane to the sphere at B with the line OA. Likewise, extend OC to OC'. This means that the triangles $\Delta\, OBC'$ and $\Delta\, OBA'$ are right triangles with right angle at B. Additionally $\angle A'B\,C' = \angle ABC = \angle B$.

Now, we have other right angles as well. First, planes $A'BC'$ and OBC are perpendicular, from above. Now, plane OCB is perpendiculare to plane OCA since $\angle ACB$ is a right angle. Look at the two planes $A'BC'$ and OAC. They intersect in the line $\overleftrightarrow{A'\,C'}$ Since $\overleftrightarrow{A'\,C'}$ lies in the plane OCA, and $\overleftrightarrow{O'\,C'}$ lies in the plane OCB, the two lines must be perpendicular. Thus, $\Delta\, O\,C'\,A'$ is a right triangle with right angle at C'.

The classical Pythagorean theorem gives:

$$(OA')^2 = (BA')^2 + (OB)^2 = (BA')^2 + R^2$$

$$(OC')^2 = (BC')^2 + (OB)^2 = (BC')^2 + R^2$$

$$(OA')^2 = (A'C')^2 + (OC')^2.$$

Combining these equations gives us:

$$(BA')^2 = (A'C')^2 + (BC')^2,$$

And so $\Delta A'B'C'$ is a triangle with right angle at C'. Let α, β and γ denote the angles at the centre of the sphere determined by the arcs BC, AC, and AB, respectively.

The definition of the sine function gives us:

$$\sin(\beta) = \frac{A'C'}{OA'} = \frac{A'C'}{BA'}\frac{BA'}{OA'} = \sin(\angle B)\sin(\gamma).$$

And so,

$$\frac{\sin\frac{b}{R}}{\sin(\angle B)} = \sin(\gamma),$$

We can reverse the roles of A and B and this gives,

$$\frac{\sin\frac{\alpha}{R}}{\sin(\angle A)} = \sin(\gamma),$$

And so,

$$\frac{\sin\frac{\alpha}{R}}{\sin(\angle A)} = \frac{\sin\frac{b}{R}}{\sin(\angle B)}.$$

Now, if we have an arbitrary triangle, we can construct the spherical analogue of the altitude to reduce the relation for two sides to the case of a right triangle, which we just finished. For example in the following figure, inserting the altitudes from A and from B, we get,

$$\sin\left(\frac{a}{R}\right)\sin(\angle C) = \sin\left(\frac{p}{R}\right) = \sin\left(\frac{c}{R}\right)\sin(\angle A).$$

From the other altitude we have,

$$\sin\left(\frac{b}{R}\right)\sin(\angle C) = \sin\left(\frac{q}{R}\right) = \sin\left(\frac{c}{R}\right)\sin(\pi - \angle B) =$$

$$\sin\left(\frac{c}{R}\right)\ \sin\ (\angle B),$$

And the theorem follows. These theorems show that there are exactly six related pieces of information determining a spherical triangle — the three sides and the three angles. Given any three of these pieces you can determine the other three pieces. This means that similar triangles must be congruent! This is not immediate. We need one other piece which we are able to prove, but will not do so now.

Theorem: Suppose that ΔABC is a right triangle on the sphere of radius R with a right angle at C. Then

$$\cos\ (A) = \sin\ (B).\ \cos\left(\frac{a}{R}\right)$$

Exercise: Small Triangles on Large Spheres

Let's look again at the formula for the sum of the angles of a spherical triangle.

$$r + g + b = \pi + \frac{1}{R^2}\text{area}(T)\ .$$

When the radius R is very large, and the area of the triangle is small, the last term on the right hand side is extremely small. In such cases it would be difficult to distinguish the sum of the angles from 180 degrees. Thus the planar sum of the angles formula is almost true, perhaps to the limits of our capability to measure angles. For example, the radius of the earth is close to 4000 miles. If we are looking at a triangle on the earth with area of 1 sq. mile, how much will the sum of the angles differ from 180 degrees? Do you think it likely that such a difference will be detectable?

Exercise: Spherical Ploygons

Girard's theorem can easily be extended from triangles to spherical polygons. Of course a spherical polygon is a figure on the sphere which is bounded by segments of great circles. Suppose that P is a spherical quadrilateral with angles a, b, c, and d.

Show that,

$$a+b+c+d=2\pi+\frac{1}{R^2}\text{area}(P).$$

Suppose that P is a spherical polygon with n sides. Show that the sum of the angles is equal to,

$$(n-2)\pi+\frac{1}{R^2}\text{area}(P).$$

Does the result for polygons apply when the polygon is a lune?

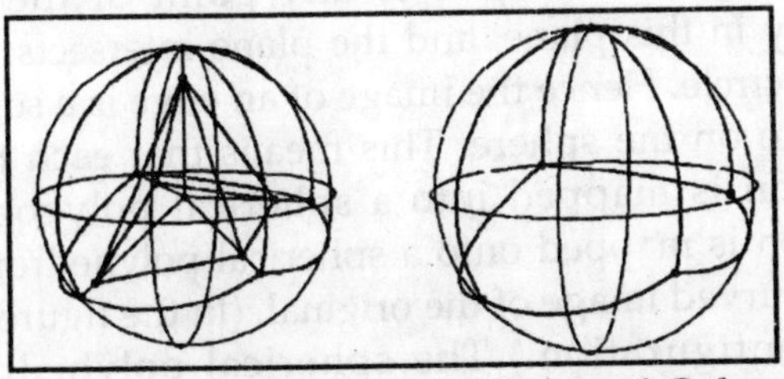

There are some states, such as Utah and Colorado, which appear to be spherical polygons. Are they really? Look at these states on a map, and measure the angle sum. Compare it with the above formula, and decide whether they are actually spherical polygons. Can you explain the discrepancy?

A PROOF OF EULER'S FORMULA

We will use Girard's Theorem, and its extenstion to spherical polygons to derive and prove and prove a famous formula of Euler's. We will state his result as a theorem.

Theorem: Let P be a convex polyhedron with V vertices, E edges, and F faces. then V– E + F = 2.

We will start the proof by choosing a point C inside P. Since P is convex, the line segment joining C to any point inside the polyhedron P, or on P itself, lies entirely within P. Next we choose a radius R so large that the sphere with center C and radius R contains the polyhedron P. We will map the polyhedron onto the sphere using central projection from C. This means that for each point on the polyhedron, we take the line from C through the point, and map the point to the intersection of this line with the sphere. This is illustrated in the accompanying figure. The original polyhedron is indicated in red, and the blue lines show how the vertices are mapped to the sphere. The black lines are reference great circles on the sphere, and the purple is the image of the polyhedron. A good way to visualize the central projection is to consider what happens if we put a light at the center of the sphere. Then the result of central projection is the shadow of the polyhedron on the sphere.

It is important to understand what happens to an edge of the polyhedron under central projection. An edge is a segment of a line. That line and the center determine a unique plane. The line segment from C to any point of the edge lies completely in this plane, and the plane intersects the sphere in a great circle. Hence the image of an edge is a segment of a great circle on the sphere. This means that each face of the polyhedron is mapped into a spherical polygon, and the polyhedron is mapped onto a spherical polyhedron which is simply a curved image of the original. (In the figure this is the purple configuration.) The spherical polyhedron has V vertices, E edges and F faces, just like P does. Furthermore, since the center of the sphere was chosen inside P, the spherical polyhedron covers the entire sphere. Hence the spherical polyhedron is a division of the sphere into F disjoint spherical polygons, which we will call Q1,..., QF.

The second figure shows the marking of the sphere determined by the polyhedron in the first figure. Let's apply Girard's Theorem to the polygon Q_i. Actually we will use the extension of Girard's Theorem to spherical polygons. Let e_i denote the number of sides of Q_i.

Then sum of angles of $Q_i = (e_i - 2)\pi + \text{area}(Q_i)/R^2$

Summing this over the faces we get:

$$\sum_{i=1}^{F}$$

sum of angles of,

$$Q_i = \sum_{i=1}^{F} (e_i\text{-}2)\pi + \sum_{i=1}^{F} \text{area (Qi)}/R^2$$

We will examine each of these sums. In each case we will be able to find the sum by geometric means. As complicated as it looks, the first sum is just the sum of all of the angles in the spherical polyhedron. Let's reorder the sum. Instead of grouping the angles by the face they belong to, let's group them by their vertex. Since the spherical polyhedron covers the sphere, at any vertex the angles with that vertex fill out the entire 2π radians. Thus the angles at each vertex contribute 2π to the sum, and multiplying by the number of vertices we see that the first sum is equal to $2\pi V$. We split the second sum into two sums.

$$\sum_{i=1}^{F} (e_i-2)\pi = \sum_{i=1}^{F} e_i \pi - \sum_{i=1}^{F} 2\pi$$

The first sum is π times the total number of all of the edges of all of the faces. Notice that each edge of the spherical poyhedron separates two faces. Since we are summing over the faces, each edge of the polyhedron is counted twice in this sum.

Therefore:

$$\sum_{i=1}^{F} e_i \pi = 2\pi E.$$

The second sum is simply,

$$\sum_{i=1}^{F} 2\pi = 2\pi F$$

Finally, since the polygons are disjoint and cover the entire sphere we have,

$$\sum_{i=1}^{F} \text{area}(Q_i)/R^2 = \text{area}(S)/R^2 = 4\pi.$$

Putting this all together we get,

$$2\pi V = 2\pi E - 2\pi F + 4\pi.$$

Dividing by 2π, and rearranging we get Euler's formula,

$$V - E + F = 2.$$

4

Geometry in Three Dimensional

INTRODUCTION

In mathematics, Cartesian geometry (analytic geometry) describes every point in three-dimensional space by means of three coordinates. Three Coordinate axes are given, each perpendicular to the other two at the origin, the point at which they cross. They are usually labeled x, y, and z. Relative to these axes, the position of any point in three-dimensional space is given by an ordered triple of real numbers, each number giving the distance of that point from the origin measured along the given axis, which is equal to the distance of that point from the plane determined by the other two axes.

Other popular methods of describing the location of a point in three-dimensional space include cylindrical coordinates and spherical coordinates, though there are an infinite number of possible methods.

Another mathematical way of viewing three-dimensional space is found in linear algebra, where the idea of independence is crucial. Space has three dimensions because the length of a box is independent of its width or breadth. In the technical language of linear algebra, space is three dimensional because every point in space can be described by a linear combination of three independent vectors. In this view, space-time is four dimensional because the location of a point in time is independent of its location in space.

Three-dimensional space has a number of properties that distinguish it from spaces of other dimension numbers. For example, at least 3 dimensions are required to tie a knot in a piece of string. Many of the laws of physics, such as the various inverse square laws, depend on dimension three.

The understanding of three-dimensional space in humans is thought to be learned during infancy using unconscious inference, and is closely related to hand-eye coordination. The visual ability to perceive the world in three dimensions is called depth perception.

SPACE

Three-dimensional space is a geometric model of the physical universe in which we live. The three dimensions are commonly called length, width, and depth (or height), although any three mutually perpendicular directions can serve as the three dimensions.

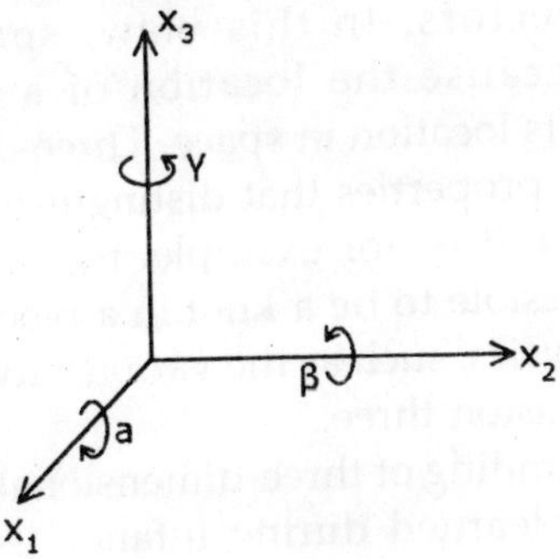

A three-dimensional picture, as one sees with a stereopticon or a View Master, tricks the human eye into experiencing the illusion of realistic depth, while the real depth is a mere fraction of a millimeter. In physics, our three-dimensional space is viewed as embedded in 4-dimensional space-time, called Minkowski space. The idea behind space-time is that time is hyperbolic-orthogonal to each of the three spatial dimensions. In mathematics, Cartesian geometry (analytic geometry) describes every point in three-dimensional space by means of three coordinates. Three Coordinate axes are given, each perpendicular to the other two at the origin, the point at which they cross.

They are usually labeled *x*, *y*, and *z*. Relative to these axes, the position of any point in three-dimensional space is given by an ordered triple of real numbers, each number giving the distance of that point from the origin measured along the given axis, which is equal to the distance of that point from the plane determined by the other two axes. Other popular methods of describing the location of a point in three-dimensional space include cylindrical coordinates and spherical coordinates, though there are an infinite number of possible methods. See Euclidean space. Another mathematical way of viewing three-dimensional space is found in linear algebra, where the idea of independence is crucial.

Space has three dimensions because the length of a box is independent of its width or breadth. In the technical language of linear algebra, space is three dimensional because every point in space can be described by a linear combination of three independent vectors. In this view, space-time is four dimensional because the location of a point in time is independent of its location in space. Three-dimensional space has a number of properties that distinguish it from spaces of other dimensions. It is, for example, the smallest dimension in which it is possible to tie a knot in a piece of string. Many of the laws of physics, such as the various inverse square laws, depend on dimension three.

The understanding of three-dimensional space in humans is thought to be learned during infancy using unconscious inference, and is closely related to hand-eye coordination. The visual ability to perceive the world in three dimensions is called depth perception. Various geometrical figures in three dimensions in three-dimensional space can be described relative to a set of mutually orthogonal axes Ox, Oy, Oz, and a point can be represented by a set of rectangular coordinates (x, y, z). The point can also be represented by cylindrical coordinates (ρ, ϕ, z) or spherical coordinates (r, θ, ϕ). In this chapter, we are concerned mostly with (x, y, z). The rectangular axes are usually chosen so that when you look down the z-axis towards the xy-plane, the y-axis is 90° counter clockwise from the x-axis. Such a set is called a right-handed set.

A left-handed set is possible, and may be useful under some circumstances, but, unless stated otherwise, it is assumed that the axes chosen in this chapter are right-handed. An equation connecting x, y and z, such as,

$$f(x,y,z) = 0 \text{ or}$$

$$z = z(x, y)$$

describes a two-dimensional surface in three-dimensional space. Aline (which need be neither straight nor two-dimensional) can be described as the intersection of two surfaces, and hence a line or curve in three-dimensional coordinate geometry is described by two equations, such as,

$$f(x, y, z) = 0 \text{ and}$$

$$g(x, y, z) = 0.$$

In two-dimensional geometry, a single equation describes some sort of a plane curve.

For example,

$$y^2 = 4qx$$

describes a parabola. But a plane curve can also be described in parametric form by two equations. Thus, a parabola can also be described by,

$$x = qt^2 \text{ and } y = 2qt$$

Similarly, in three-dimensional geometry, a line or curve can be described by three equations in parametric form. For example, the three equations,

$$x = a \cos t$$

$$y = a \sin t$$

$$z = ct$$

describe a curve in three-space. Think of the parameter t as time, and see if you can imagine what sort of a curve this is.

We shall be concerned in this chapter mainly with six types of surface: the plane, the ellipsoid, the paraboloid, the hyperboloid, the cylinder and the cone.

ELLIPSOID

A long ellipsoid object looks like a cigar; a short one looks like somewhat like a pill. Its cross sections are circles or ellipses. The elllipsoid object has the same attributes as the box object and it can be thought of as fitting inside a box of the same dimensions: myell = ellipsoid(pos=(x_0,y_0,z_0), length=L, height=H, width=W) The given position is in the center of the ellipsoid, at (x_0, y_0, z_0). This is different from cylinder, whose pos attribute is at one end of the cylinder. Just as with a cylinder, we can refer to the individual vector components of the ellipsoid as myell.x, myell.y, and myell.z. The length from end to end (along the x axis) is L, the height (along the y axis) is H, and the width is W (along the z axis). For this ellipsoid, we have myell.axis = (L, 0, 0). Note that the axis of an ellipsoid is just like the axis of a cylinder.

For an ellipsoid that isn't aligned with the coordinate axes, additional issues come into play. The orientation of the length of the ellipsoid is given by the axis: myell = ellipsoid (pos=(x_0,y_0,z_0), axis=(a,b,c), length=L, height = H, width = W)

The axis attribute gives a direction for the length of the ellipsoid, and the length, height, and width of the ellipsoid are given as before (if a length attribute is not given, the length is set to the magnitude of the axis vector).

The ellipsoid object has the following attributes and default values, like those for cylinders: pos (0,0,0), x (0), y(0), z(0), axis (1,0,0), length (1), colour (1,1,1) which is colour.white, red (1), green (1), blue (1), and up (0,1,0).

Additional box attributes: height In the y direction in the simple case, default is 1. Width In the z direction in the simple case, default is 1. size (length, height, width), default is (1,1,1) myell.size=(20,10,12) sets length=20, height=10, width=12

Note that the pos attribute for cylinder, arrow, cone, and pyramid corresponds to one end of the object, whereas for an ellipsoid, box, sphere, or ring it corresponds to the center of the object.

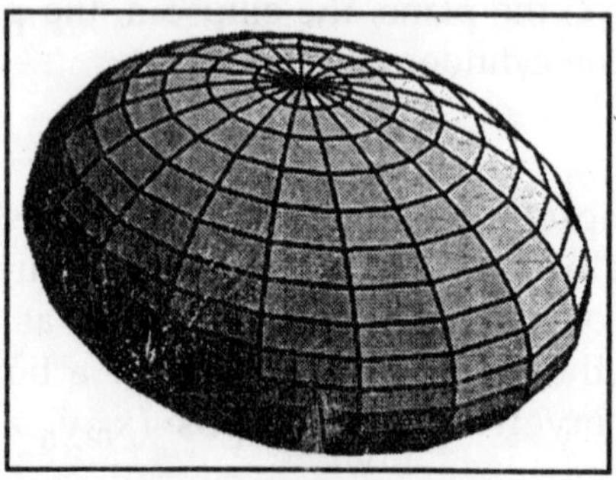

The general ellipsoid, also called a triaxial ellipsoid, is a quadratic surface which is given in Cartesian coordinates by,

$$\frac{x^2}{a^2}+\frac{y^2}{b^2}+\frac{z^2}{c^2}=1, \tag{1}$$

where the semi-axes are of lengths a, b, and c. In spherical coordinates, this becomes,

$$\frac{r^2\cos^2\theta\sin^2\phi}{a^2}+\frac{r^2\sin^2\theta\sin^2\phi}{b^2}+\frac{r^2\cos^2\phi}{c^2}=1. \tag{2}$$

If the lengths of two axes of an ellipsoid are the same, the figure is called a spheroid (depending on whether c<a orc > a, an oblate spheroid or prolate spheroid, respectively), and if all three are the same, it is a sphere. Tietze calls the general ellipsoid a "triaxial ellipsoid." There are two families of parallel circular cross sections in every ellipsoid. However, the two coincide for spheroids. If the two sets of circles are fastened together by suitably chosen slits so that they are free to rotate without sliding, the model is movable. Furthermore, the disks can always be moved into the shape of a sphere.

In 1882, Staude discovered a "thread" construction for an ellipsoid analogous to the taut pencil and string construction of the ellipse. This construction makes use of a fixed framework consisting of an ellipse and a hyperbola. The parametric equations of an ellipsoid can be written as,

$$x = a\cos u\sin v$$

$$y = b\sin u\sin v$$

$$c\cos v.$$

for $u\in[0,2\pi)$ and $v\in[0,\pi]$. In this parametrization, the coefficients of the first fundamental form are

$$F = (b^2\cos^2 u+a^2\sin^2 u)\sin^2 v$$

$$(b^2-a^2)\cos u\sin u\cos v\sin v$$

$$(a^2\cos^2 u+b^2\sin^2 u)\cos^2 v+c^2\sin^2 v.$$

and of the second fundamental form are,

$$e=\frac{abc\sin^2 v}{\sqrt{a^2b^2\cos^2 v+c^2(b^2\cos^2 u+a^2\sin^2 u)\sin^2 v}}$$

$$0$$

$$\frac{abc}{\sqrt{a^2b^2\cos^2 v + c^2(b^2\cos^2 u + a^2\sin^2 u)\sin^2 v}}$$

Also in this parametrization, the Gaussian curvature is:

$$K = \frac{a^2 b^2 c^2}{[a^2b^2\cos^2 v + c^2(b^2\cos^2 u + a^2\sin^2 u)\sin^2 v]^2}$$

and the mean curvature is,

$$H = \frac{abc[3(a^2+b^2)+2c^2+(a^2+b^2-2c^2)\cos(2v)-2(a^2-b^2)\cos(2u)\sin^2 v]}{8[a^2b^2\cos^2 v + c^2(b^2\cos^2 u + a^2\sin^2 u)\sin^2 v]^{3/2}}$$

The Gaussian curvature can be given implicitly by,

$$\frac{a^2b^6c^6}{[c^4b^4 + c^4(b^2-b^2)y^2 + b^4(a^2-c^2)z^2]^2}$$

$$\frac{a^6b^6c^2}{[a^4c^4 + a^4(b^2-c^2)z^2 + c^4(b^2-a^2)x^2]^2}$$

$$\frac{a^6b^6c^2}{[a^4b^4 + b^4(c^2-a^2)x^2 + a^4(c^2-b^2)y^2]^2}.$$

The surface area of an ellipsoid is given by,

$$2\pi a\, b\, \mathrm{ns}\theta \int_0^{\theta} \left(\frac{\mathrm{dn}^2\theta}{\mathrm{dn}^2 u} + \frac{\mathrm{cn}^2\theta}{\mathrm{cn}^2 u} \right) du$$

$$2\pi \left[c^2 + \frac{bc^2}{\sqrt{a^2-c^2}}\theta + b\sqrt{a^2-c^2}\,E(\mathrm{am}(\theta), k) \right].$$

where ns (θ), dn (θ), and (θ) are Jacobi elliptic functions with modulus k,

$$k \equiv \frac{e_2}{e_1}$$

$$e_1 \equiv \sqrt{\frac{a^2-c^2}{a^2}}$$

$$e_2 \equiv \sqrt{\frac{b^2 - c^2}{b^2}}.$$

$E(\phi, k)$ is an incomplete elliptic integral of the second kind, $am(\phi)$ is the Jacobi amplitude with modulus k, and θ is given by inverting the expression,

$$e_1 = sn(\theta, k),$$

where $sn(\theta)$ is another Jacobi elliptic function with modulus k; error corrected). Another form of the surface area equation is,

$$S = 2\pi\left[c^2 + \frac{bc^2}{\sqrt{a^2 - c^2}}F(\phi, k) + b\sqrt{a^2 - c^2}E(\phi, k)\right],$$

where,

$$\phi = \sin^{-1}\left(\sqrt{1 - \frac{c^2}{a^2}}\right)$$

The surface area can also be obtained directly from the first fundamental form as,

$$\int_0^n \int_0^{2n} \sqrt{EG - F^2}\, d\theta d\phi$$

$$\int_0^n \sin\phi \int_0^{2n} \sqrt{a^2b^2\cos^2\phi + c^2(b^2\cos^2\theta + a^2\sin^2\theta)\sin^2\phi}\, d\theta d\phi$$

$$2\sqrt{2}\, b\int_0^n \sqrt{a^2 + c^2 + (a^2 - c^2)\cos(2\phi)}\, \sin\phi \times E$$

$$\left(\frac{c}{b}\sqrt{\frac{2(b^2 - a^2)}{a^2 + c^2 + (a^2 - c^2)\cos(2\phi)}}\sin\phi\right)d\phi$$

The volume of an ellipsoid is,

$$V = \frac{4}{3}\pi abc.$$

The geometric centroids of the half-ellipsoids along the x–, y–, and z–axes are, $\bar{x} = \frac{3}{16}a$.

$$\bar{y} = \frac{3}{16}b$$

$$\bar{z} = \frac{3}{16}c.$$

The moment of inertia tensor of an ellipsoid is given by,

$$I = \begin{bmatrix} \frac{1}{5}m(b^2+c^2) & 0 & 0 \\ 0 & \frac{1}{5}M(a^2+c^2) & 0 \\ 0 & 0 & \frac{1}{5}M(a^2+c^2) \end{bmatrix}$$

A different parameterization of the ellipsoid is the so-called stereographic ellipsoid, given by the parametric equations,

$$x(u,v) = \frac{a(1-u^2-v^2)}{1+u^2+v^2}$$

$$y(u,v) = \frac{2bu}{1+u^2+v^2}$$

$$z(u,v) = \frac{2cv}{1+u^2+v^2}$$

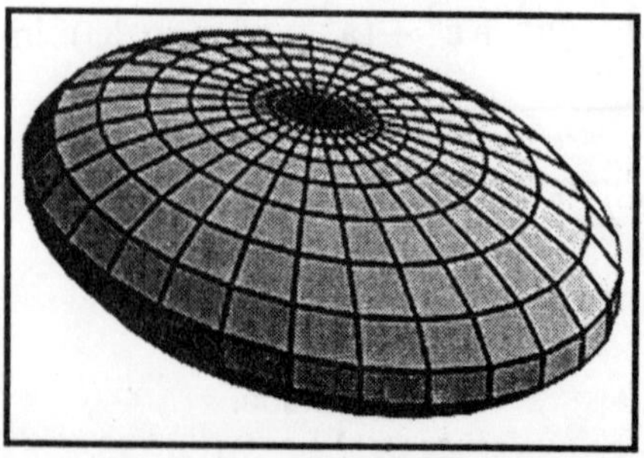

A third parameterization is the Mercator parameterization,

$$x(u, v) = a \operatorname{sech} v \cos u$$

$$y(u, v) = b \operatorname{sech} v \sin u$$

$$z(u,v) = c \tanh v$$

The support function of the ellipsoid is,

$$h = \left(\frac{x^2}{a^4} + \frac{y^2}{b^4} + \frac{z^2}{C^2} \right)^{-1/2},$$

and the Gaussian curvature is,

$$K = \frac{h^4}{a^2 b^2 c^2}$$

THREE DIMENSION PLANES AND STRAIGHT LINES

The geometry of the plane and the straight line is, of course, rather simple, so that we can dispose of them in this brief introductory section as.

The equation,

$$Ax + By + Cz + D = 0$$

represents a plane. If $D \neq 0$ it is often convenient, and saves algebra and computation with no loss of information, to divide the equation through by D and re-write it in the form,

$$ax + by + cz = 1.$$

The coefficients need not by any means all be positive. If $D = 0$, the plane passes through the origin of coordinates, and it may be convenient to divide the equation by C and hence to rewrite it in the form,

$$ax + by + z = 0.$$

The plane represented by equation intersects the yz–, zx– and xy–planes in the straight lines,

$$by + cz = 1$$

$$cz + ax = 1$$

$$ax + by = 1$$

and it intersects the x-, y- and z-axes at,

$$x = x_0 = 1/a$$

$$y = y_0 = 1/b$$

$$z = z_0 = 1/c$$

The geometry can be seen in figure.

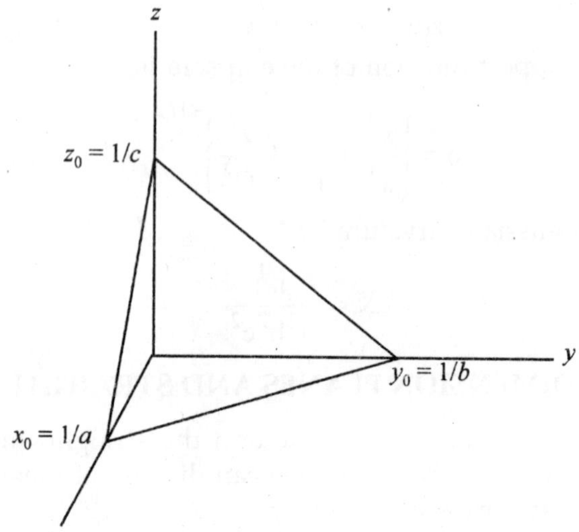

Another way of writing the equation to the plane would be,

$$\frac{x}{x_0} + \frac{y}{y_0} + \frac{z}{z_0} = 1.$$

In this form, x_0, y_0 and z_0 are the intercepts on the x-, y- and z-axes. Let $P(x, y, z)$ be a point on the plane 4.2.2. The square S of the distance of that point from the origin is,

$$S = x^2 + y^2 + z^2.$$

But since (x, y, z) is on the plane, we can write S in terms of x and y alone, by substituting for z from equation:

$$S = x^2 + y^2 + \frac{(1 - ax - by)^2}{c^2}.$$

The distance to the origin is least for a point on the plane such that $\frac{\partial S}{\partial x}$ and $\frac{\partial S}{\partial y}$ are both zero. After some algebra, we find that the point on the plane that is closest to the origin is the point,

$$\left(\frac{a}{a^2 + b^2 + c^2}, \frac{b}{a^2 + b^2 + c^2}, \frac{c}{a^2 + b^2 + c^2}\right)$$

Consequently the perpendicular distance of the origin

from the plane is the orthogonal sum of these coordinates, which is,

$$\frac{1}{\sqrt{a^2+b^2+c^2}}$$

Substitution of x_0 for $1/a$, y_0 for $1/b$ and z_0 for $1/c$ will make these results seem rather obvious. Further, *OP* is normal to the plane, and the *direction cosines* of *OP*, i.e. of the normal to the plane, are,

$$\frac{a}{\sqrt{a^2+b^2+c^2}}, \frac{b}{\sqrt{a^2+b^2+c^2}}, \frac{c}{\sqrt{a^2+b^2+c^2}}$$

The coefficients *a*, *b*, *c* are direction ratios of the normal to the plane; that is to say, they are numbers that are proportional to the direction cosines.

Example: Consider the plane, $0.50x + 0.25y + 0.20z = 1$

The plane intersects the *x*-, *y*- and *z*-axes at (2, 0, 0), 0, 4, 0) and (0, 0, 5). The point on the plane that is closest to the origin is (1.4184, 0.7092, 0.5674). The perpendicular distance of the origin from the plane is 1.6843. The direction cosines of the normal to the plane are (0.8422, 0.4211, 0.3369). Now, let P(*x*, *y*, *z*) be a point in the plane and let P_0 (x_0, y_0, z_0) be some point not necessarily in the plane. It is of interest to find the perpendicular distance between P_0 and the plane. Let *S* be the square of the distance between P_0 and *P*.

Then,

$$S = (x - x_0)^2 + (y - y_0)^2 + (z - z_0)^2$$

We can express this in terms of *x* and *y* alone by substitution for *z* from equation. The distance of *P* from P_0 is least when $\frac{\partial S}{\partial x}$ and $\frac{\partial S}{\partial y}$ are simultaneously zero. These two conditions result in,

$$(a^2 + c^2)x = a + c^2 x_0 - acz_0 - aby$$

$$(b^2 + c^2)y = b + c^2 y_0 - bcz_0 - abx$$

These, combined with equation, result in,

$$x = \frac{(b^2+c^2)x_0 + a(1 - by_0 - cz_0)}{a^2+b^2+c^2}$$

$$y = \frac{(c^2 + a^2)y_0 + b(1 - cz_0 - ax_0)}{a^2 + b^2 + c^2}$$

$$z = \frac{(a^2 + b^2)z_0 + c(1 - ax_0 - by_0)}{a^2 + b^2 + c^2}$$

These are the coordinates of the point P in the plane that is nearest to P_0. The perpendicular distance between P and P_0 is $p = \sqrt{S}$, or,

$$p = \frac{1 - ax_0 - by_0 - cz_0}{\sqrt{a^2 + b^2 + c^2}}.$$

This is positive if P_0 is on the same side of the plane as the origin, and negative if it is on the opposite side. If the perpendicular distances of two points from the plane, as calculated from equation, are of opposite signs, they are on opposite sides of the plane. If $p = 0$, or indeed if the numerator of equation is zero, the point P_0 (x_0, y_0, z_0) is, of course, in the plane. An equation for the plane containing three specified points can be found as follows. Let (x_1, y_1), (x_2, y_2), (x_3, y_3) be the three specified points, and let (x, y) be any point in the plane that contains these three points. Each of these points must satisfy an equation of the form.

That is,

$$xA + yB + zC + D = 0$$
$$x_1A + y_1B + z_1C + D = 0$$
$$x_2A + y_2B + z_2C + D = 0$$
$$x_3A + y_3B + z_3C + D = 0$$

In these equations, we are treating A, B, C, D as unknowns, and the x, y, z, x_1, y_1... as coefficients. We have four linear equations in four unknowns, and no constant term. From the theory of equations, these are consistent only if each is a linear combination of the other three. This is satisfied only if the determinant of the coefficients is zero:

$$\begin{vmatrix} x & y & z & 1 \\ x_1 & y_1 & z_1 & 1 \\ x_2 & y_2 & z_2 & 1 \\ x_3 & y_3 & z_3 & 1 \end{vmatrix} = 0$$

and this is the equation to the required plane containing the three points. The reader will notice the similarity of this equation to equation for a line passing between two points in twodimensional geometry. The reader might like to repeat the argument, but requiring instead the four points to satisfy an equation of the form. There will then be four linear equations in three unknowns. Otherwise the argument is the same.

We now move on to the question of finding the area of a triangle whose vertices are given. It is straightforward to do this with a numerical example, and the reader is now encouraged to write a computer programme, in whatever language is most familiar, to carry out the following tasks. Read as data the *x-y-z* coordinates of three points *A, B, C.*

Calculate the lengths of the sides *a, b, c, a* being opposite to A, etc. Calculate the three angles at the vertices of the triangle, in degrees and minutes, and check for correctness by verifying that their sum is 180°. If an angle is obtuse, make sure that the computer displays its value as a positive angle between 90° and 180^0. Finally, calculate the area of the triangle.

It should be easy to calculate the sides. The angles can then be calculated from equation and the area from each of the four equations. They should all yield the correct answer, of course, but the redundant calculations serve as an important check on the correctness of your programming, as also does your check that the three angles add to 180°. Where there are two of more ways of performing a calculation, a careful calculator will do all of them as a check against mistakes, whether the calculation is done by hand or by computer.

Example: If the coordinates of the vertices are, A(7, 4, 3), B(11, 6, 2), C(9, 2, 4) the sides are $a = 4.583$, $b = 3.000$, $c = 4.899$, and the angles are A = 36° 42', B = 77° 24', C = 65° 54', which add up to 180°. The area is 6.708.

Example: If the coordinates of the vertices are,

A(6, 4, 9), B(2, 6, 17), C(8, 3, 5)

the area of the triangle is zero and the points are collinear. The foregoing showed that it was not difficult to calculate numerically the area of the triangle from the coordinates of its vertices. Is it easy to find a simple explicit algebraic formula for the area in terms of (x_1, y_1, z_1), (x_2, y_2, z_2) and (x_3, y_3, z_3)? On referring to figure, we can proceed as follows.

The vectors r_2 and r_3 can be written,

$$r_2 = (x_2 - x_1)i + (y_2 - y_1)j + (z_2 - z_1)k$$
$$r_3 = (x_3 - x_1)i + (y_3 - y_1)j + (z_3 - z_1)k$$

where i, j, k are unit vectors parallel to the x-, y- and z- axes. The cross product of r_2 and r_3 gives the (vector) area of the parallelogram of which they form two sides. The area A of the triangle is half of this, so that,

$$\begin{aligned} 2A &= r_2 \times r_3 \\ &= [(y_2 - y_1)(z_3 - z_1) - (y_3 - y_1)(z_2 - z_1)]i \\ &+ [(z_2 - z_1)(x_3 - x_1) - (z_3 - z_1)(x_2 - x_1)]j \\ &+ [(x_2 - x_1)(y_3 - y_1) - (x_3 - x_1)(y_2 - y_1)]k \end{aligned}$$

The magnitude of this vector can be found in the usual way, to obtain,

$$\begin{aligned} 4A^2 &= [(y_1 (z_2 - z_3) + y_2 (z_3 - z_1) + y_3 (z_1 - z_2)]^2 \\ &+ [(z_1 (x_2 - x_3) + z_2 (x_3 - x_1) + z_3 (x_1 - x_2)]^2 \\ &+ [(x_1(y_2 - y_3) + x_2 (y_3 - y_1) + x_3 (y_1 - y_2)]^2. \end{aligned}$$

The reader can verify that, if $z_1 = z_2 = z_3$, this reduces to equation for the area of a triangle the xy-plane. Equation can also be written,

$$4A^2 = \begin{vmatrix} y_1 & y_2 & y_3 \\ z_1 & z_2 & z_3 \\ 1 & 1 & 1 \end{vmatrix}^2 + \begin{vmatrix} z_1 & z_2 & z_3 \\ x_1 & x_2 & x_3 \\ 1 & 1 & 1 \end{vmatrix}^2 + \begin{vmatrix} x_1 & x_2 & x_3 \\ y_1 & y_2 & y_3 \\ 1 & 1 & 1 \end{vmatrix}^2$$

This gives the area explicitly in terms of the coordinates of the vertices. If it is zero, the points are collinear.

The volume of a tetrahedron is, $\frac{1}{6}$ × base × height. By combining equation for the area of a triangle with equation for the perpendicular distance of a point form a plane, we can determine that the volume of the tetrahedron whose vertices are, (x_0, y_0, z_0), (x_1, y_1, z_1), (x_2, y_2, z_2), (x_3, y_3, z_3) is,

$$\frac{1}{6}\begin{vmatrix} x_1 & y_1 & z_1 & 1 \\ x_2 & y_2 & z_2 & 1 \\ x_3 & y_3 & z_3 & 1 \end{vmatrix}$$

If this determinant is zero, the four points are coplanar. In three-dimensional coordinate geometry, a straight line is described by two equations, being the intersection of two planes:

$$a_1x + b_1x + c_1z = 1$$

$$a_2x + b_2x + c_2z = 1$$

If $a_1/a_2 = b_1/b_2 = c_1/c_2$, the normals to the two planes have the same direction ratios, so the planes are parallel and do not intersect. Otherwise the normals to the two planes have different direction ratios (a_1, b_1, c_1), (a_2, b_2, c_2), and, since the line of intersection of the planes is at right angles to both normals, the direction ratios of the line are found from the cross product of vectors normal to the planes. The direction ratios of the line of intersection are therefore

$$(b_1c_2 - b_2c_1, c_1a_2 - c_2a_1, a_1b_2 - a_2b_1)$$

The line crosses the *yz*-, *zx*- znd *xy*- planes at,

$$y = \frac{c_1 - c_2}{b_1c_2 - b_2c_1} \qquad z = \frac{b_1 - b_2}{b_1c_2 - b_2c_1}$$

$$z = \frac{a_1 - a_2}{c_1a_2 - c_2a_1} \qquad x = \frac{c_1 - c_2}{c_1a_2 - c_2a_1}$$

$$x = \frac{b_1 - b_2}{a_1b_2 - a_2b_2} \qquad y = \frac{a_1 - a_2}{a_1b_2 - a_2b_1}$$

An example of computing a straight line from the intersection of two planes occurs in meteor astronomy. We can assume a flat Earth, which is tantamount to supposing that the height of a meteor is negligible compared with the radius of Earth, and the height of an observer above sea level is negligible compared with the height of the meteor. Since the heights of meteors are typically a few tens of km, both of these approximations are reasonable, at least for noninstrumental eyewitness accounts.

We suppose that, relative to an arbitrary origin O on the surface of Earth, a witness A is 15 km east and 5 km north of the origin. He sees a fireball start at an angle $\theta = 25°.5$ from his zenith and at an azimuth $\phi = 54°.5$ counterclockwise from his east, and it finishes at $\theta = 36°.7$, $\phi = 16°.7$.

Show that the plane containing the witness and the meteor is,

$$0.0363x + 0.0911y - 0.0454z = 1$$

A second witness, 30 km east and 15 km north of O, estimates the zenith distance and azimuth of two points on the meteor track to be $\theta = 29°.6$, $\phi = 202°.9$ and $\theta = 33°.6$, $\phi = 242°.9$ Show that the plane containing this second witness and the meteor is,

$$0.0257x + 0.0153y + 0.0168z = 1$$

These two equations describe the path of the fireball through the air. Show that, if the meteoroid carries on moving in a straight line, it will strike the ground as a meteorite 42.4 km east and 6.0 km south of the origin O. As discussed, two nonparallel planes intersect in a straight line. Usually, three nonparallel planes intersect at a single unique point; for, if L is a line formed from the intersection of planes P_1 and P_2, L will usually intersect the plane P_3 at a point.

Example: The planes,

$$2x + 3y + 4z - 9 = 0$$

$$x + y - 8z + 6 = 0$$

$$5x + 6y - 12z + 1 = 0$$

intersect at (1, 1, 1). It will be recalled from the theory of linear equations that three equations,

$$A_1x + B_1y + C_1z + D_1 = 0$$
$$A_2x + B_2y + C_2z + D_2 = 0$$
$$A_3x + B_3y + C_3z + D_3 = 0$$

have a unique solution only if,

$$\Delta = \begin{vmatrix} A_1 & B_1 & C_1 \\ A_2 & B_2 & C_2 \\ A_3 & B_2 & C_3 \end{vmatrix} \neq 0$$

and, in the geometrical interpretation, this is the condition that three planes meet in a single point.

Consider, however, the three planes,

$$2x + 3y + 4z - 9 = 0$$

$$x + y - 8z + 6 = 0$$

$$5x + 6y - 20z + 12 = 0$$

The direction ratios of the three lines found by combining the planes in pairs are,

$$(-28, 20, -1)\ (-84, 60, -3)\ (28, -20, 1)$$

It will be observed that each is a multiple of either of the others, and the direction cosines of each of the three lines are identical apart from sign: ($\mp 0.813, \pm 0.581, \mp\ 0.029$). The three lines are, in fact, parallel, and the three planes enclose a prism. A condition for this is that,

$$\Delta = 0.$$

But consider now the planes,

$$2x + 3y + 4z - 9 = 0$$

$$x + y - 8z + 6 = 0$$

$$5x + 6y - 20z + 9 = 0$$

Not only does $\Delta = 0$, but also,

$$\Delta' = \begin{vmatrix} A_1 & B_1 & D_1 \\ A_2 & B_2 & D_2 \\ A_3 & B_3 & D_3 \end{vmatrix} = 0$$

The three lines obtained by combining equations in pairs are in fact identical, and the three planes meet in a single line. Each of equations is a linear combination of the other two.

In summary, three nonparallel planes meet in a single line if $\Delta \neq 0$. They meet in a single point if $\Delta = \Delta' = 0$. They enclose a prism if $\Delta = 0$, $\Delta' \neq 0$.

ELLIPTICITY

Consider the equation,

$$\frac{x^2}{a^2} + \frac{z^2}{c^2} = 1,$$

with $a > c$, in the xz-plane. The length of the semi major axis is a and the length of the semi minor axis is c. If this figure is rotated through 360° about its minor (z-) axis, the three-dimensional figure so obtained is called an *oblate spheroid*. The figure of the Earth is not exactly spherical; it approximates to a very slightly oblate spheroid, the ellipticity $(c - a)/a$ being only 0.00335.

The equation to the oblate spheroid referred to above is,

$$\frac{x^2}{a^2} + \frac{y^2}{a^2} + \frac{z^2}{c^2} = 1 .$$

If the ellipse is rotated through 360° about its major (x-) axis, the figure so obtained is called a *prolate* spheroid. A rugby football (or, to a lesser extent, a North American football, which is a bit too pointed) is a good approximation to a prolate spheroid. The equation to the prolate spheroid just described is,

$$\frac{x^2}{a^2} + \frac{y^2}{c^2} + \frac{z^2}{c^2} = 1 .$$

Either type of spheroid can be referred to as an "ellipsoid of revolution". The figure described by the equation,

$$\frac{x^2}{a^2} + \frac{y^2}{b^2} + \frac{z^2}{c^2} = 1$$

is a *tri-axial ellipsoid*. Unless stated otherwise, I shall adopt the convention $a > b > c$, and choose the coordinate axes such that the major, intermediate and minor axes are along the x-, y- and z-axes respectively. A tri-axial ellipsoid is not an ellipsoid of revolution; it cannot be obtained by rotating an ellipse about an axis.

The special case $a = b = c$: $x^2 + y^2 + z^2 = a^2$ is, of course, a sphere.

There are actually two such directions, symmetrically situated on either side of the z-axis, but there are no such directions in either the xy- or the yz-planes from which the cross-section of the ellipsoid appears as a circle. Expressed otherwise, there are two planes that intersect the ellipsoid in a circle.

This fact is of some importance in the description of the propagation of light in a bi-axial crystal, in which one of the wavefronts is a tri-axial ellipsoid. Let us refer the ellipsoid to a set of axes $Ox'y'z'$ such that the angles $z'Oz$ and $x'Ox$ are each θ, and the y'- and y-axes are identical. The equation of the ellipsoid referred to the new axes is (by making use of the usual formulas for the rotation of axes)

$$\frac{(z'\sin\theta + x'\cos\theta)^2}{a^2} + \frac{y^2}{b^2} + \frac{(z'\cos\theta - x'\sin\theta)^2}{c^2} = 1.$$

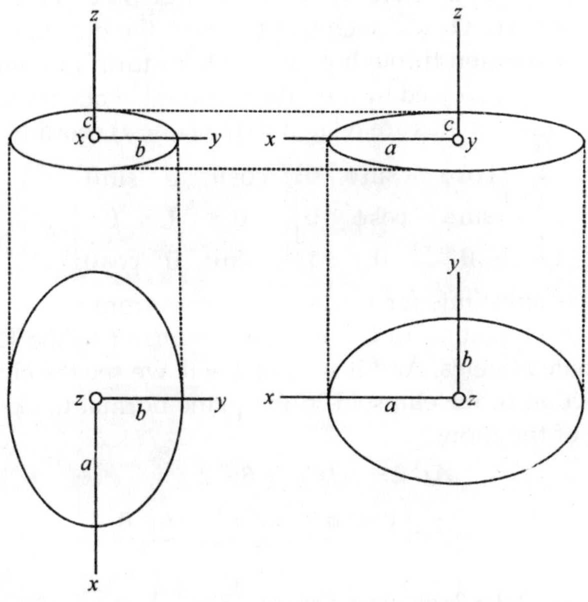

Figure shows the cross-section of a tri-axial ellipse in the yz- plane (a), the xz-plane (b) and (twice - (c), (d)) the xy-plane. If you imagine your eye wandering in the xz-plane from the x-axis (a) to the z-axis (c), you will be convinced that there is a direction in the xz-plane from which the cross-section of the ellipse is a circle.

The cross-section of the ellipsoid in the $x'y'$-plane (i.e. normal to the z'-axis) is found by putting $z' = 0$:

$$\frac{(x'\cos\theta)^2}{a^2} + \frac{y^2}{b^2} + \frac{(x'\sin\theta)^2}{c^2} = 1.$$

This is a circle if the coefficients of x' and y' are equal. Thus it is a circle if,

$$\cos^2\theta = \frac{a^2(b^2 - c^2}{b^2(a^2 - c^2)}.$$

Thus, a plane whose normal is in the xz-plane (i.e. between the major and minor axis) and inclined at an angle θ to the minor (z-) axis, cuts the tri-axial ellipsoid in a circle. As viewed from either of these directions, the cross-section of the ellipsoid is a circle of radius b. Let us erect a set of coordinate axes $Ox'y'z'$ such that Oz' is in the direction (θ, ϕ), first by a rotation through ϕ about Oz to form intermediate axes $Ox_1y_1z_1$, followed by a rotation through θ about Oy_1. The (x', y', z') coordinates are related to the (x, y, z) coordinates by

$$\begin{pmatrix} x \\ y \\ z \end{pmatrix} = \begin{pmatrix} \cos\phi & -\sin\phi & 0 \\ \sin\phi & \cos\phi & 0 \\ 0 & 0 & 1 \end{pmatrix} \begin{pmatrix} \cos\theta & 0 & \sin\theta \\ 0 & 1 & 0 \\ -\sin\theta & 0 & \cos\theta \end{pmatrix} \begin{pmatrix} x' \\ y' \\ z' \end{pmatrix}$$

If we substitute for x, y, z in equation from equation, we obtain the equation to the ellipsoid referred to the $Ox'y'z'$ coordinate systems. And if we put $z' = 0$, we see the elliptical crosssection of the ellipsoid in the plane normal to Oz'. This will be of the form,

$$Ax'2 + 2Hx'y' + By'2 = 1,$$

where,

$$A = \cos^2\theta\left(\frac{\cos^2\phi}{a^2} + \frac{\sin^2\phi}{b^2}\right) + \frac{\sin^2\theta}{c^2},$$

$$2H = 2\cos^2\theta\sin\phi\cos\phi\left(\frac{1}{b^2} - \frac{1}{a^2}\right),$$

$$B = \frac{\sin^2\phi}{a^2} + \frac{\cos^2\phi}{b^2}.$$

This is an ellipse whose axes are inclined at an angle ψ from Ox' given by,

$$\tan 2\psi = \frac{2H}{A - B}.$$

By replacing x' and y' by x'' and y''. where,

$$\begin{pmatrix} x' \\ y' \end{pmatrix} = \begin{pmatrix} \cos\psi & -\sin\psi \\ \sin\psi & \cos\psi \end{pmatrix} \begin{pmatrix} x'' \\ y'' \end{pmatrix}$$

we shall be able to describe the ellipse in a coordinate system $Ox''y''$ whose axes are along the axes of the ellipse, and the equation will be of the form

$$\frac{x''^2}{a''^2}+\frac{y''^2}{b''^2}=1$$

and the area of the cross-section is $\pi a''b''$. For example, suppose the semi axes of the ellipsoid are $a = 3$, $b = 2$, $y = 1$, and we look at it from the direction $\theta = 60°$, $\phi = 45°$. Following equations we obtain for the equation of the elliptical cross-section referred to the system $Ox'y'z'$ $0.79513\dot{8}\ x'^2 + 0.069\dot{4}\ x'\ y' + 0.180\dot{5}\ y'2 = 1$. From equation we find $\psi = 3°.22338$. Equation then transforms equation to $0.797094x''^2 + 0.178600y''^2 = 1$ or,

$$\frac{x''^2}{(1.1201)^2}+\frac{y''^2}{(2.3662)^2}=1.$$

The area is $\pi \times 1.1202 \times 2.3663 = 8.362$.

PARABOLOID

The equation $x = 4qz = 2lz$ is a parabola in the xz-plane. The distance between vertex and focus is q, and the length of the semi latus rectum $l = 2q$.

The equation can also be written,

$$\frac{x^2}{a^2}=\frac{z}{h}$$

Here a and h are distances such that $x = a$ when $z = h$, and the length of the semi latus rectum is $l = a^2/(2h)$. If this parabola is rotated through 360° about the z-axis, the figure swept out is a *paraboloid of revolution*, or *circular paraboloid*. Many telescope mirrors are of this shape.

The equation to the circular paraboloid is

$$\frac{x^2}{a^2}+\frac{y^2}{a^2}=\frac{z}{h}.$$

The cross-section at $z = h$ is a circle of radius a. The equation,

$$\frac{x^2}{a^2}+\frac{y^2}{a^2}=\frac{z}{h},$$

in which we shall choose the x- and y-axes such that $a > b$, is an elliptic paraboloid and, if $a \neq b$, is not formed by rotation of a parabola. At $z = h$, the cross section is an ellipse of semi major and minor axes equal to a and b respectively. The section in the plane $y = 0$ is a parabola of semi latus rectum $a^2/(2h)$. The section in the plane $x = 0$ is a parabola of semi latus rectum $b^2/(2h)$. The elliptic paraboloid lies entirely above the xy-plane.

The equation,

$$\frac{x^2}{a^2} - \frac{y^2}{b^2} = \frac{z}{h}$$

is a hyperbolic paraboloid, and its shape is not quite so easily visualized. Unlike the elliptic paraboloid, it extends above and below the plane. It is a saddle-shaped surface, with the saddle point at the origin.

The section in the plane $y = 0$ is the "nose down" parabola $x^2 = a^2 z/h$ extending above the xy-plane. The section in the plane $x = 0$ is the "nose up" parabola $y^2 = -b^2z/h$ extending below the xy-plane. The section in the plane $z = h$ is the hyperbola,

$$\frac{x^2}{a^2} - \frac{y^2}{b^2} = 1.$$

The section with the plane $z = -h$ is the conjugate hyperbola,

$$\frac{x^2}{a^2} - \frac{y^2}{b^2} = -1.$$

The section with the plane $z = 0$ is the asymptotes

$$\frac{x^2}{a^2} - \frac{y^2}{b^2} = 0.$$

HYPERBOLOID

The equation,

$$\frac{x^2}{a^2} - \frac{z^2}{c^2} = 1$$

is a hyperbola, and a is the semi transverse axis.

If it is rotated about the z-axis through 360°, the surface swept out is a *circular hyperboloid* of one sheet.

Its equation is,

$$\frac{x^2}{a^2}+\frac{y^2}{a^2}-\frac{z^2}{c^2}=1.$$

Imagine two horizontal rings, one underneath the other. The upper one is fixed. The lower one is suspended from the upper one by a large number of vertical strings attached to points equally spaced around the circumference of each ring. Now twist the lower one through a few degrees about a vertical axis, so that the strings are no longer quite vertical, and the lower ring rises slightly. These strings are generators of a circular hyperboloid of one sheet.

If it is rotated about the z-axis through 360°, the surface swept out is a *circular hyperboloid of two sheets.* Its equation is,

$$\frac{x^2}{a^2}-\frac{y^2}{c^2}-\frac{z^2}{c^2}=1.$$

The equations,

$$\frac{x^2}{a^2}+\frac{y^2}{c^2}-\frac{z^2}{c^2}=1$$

and,

$$\frac{x^2}{a^2}-\frac{y^2}{c^2}-\frac{z^2}{c^2}=1$$

represent hyperbolas of one and two sheets respectively, but are not hyperbolas of revolution, since their cross sections in the planes z = constant and x = constant > a respectively are ellipses rather than circles. The reader should imagine what the cross- sections of all four hyperboloids are like in the planes $x = 0$, $y = 0$ and $z = 0$.

CYLINDER

In three-dimensional solid geometry the equation

$$\frac{x^2}{a^2}+\frac{y^3}{b^2}=1$$

represents a cylinder of elliptical cross-section, whose axis coincides with the z-axis. The equation to a cylinder with an axis in another position and with another orientation can be obtained by the usual processes of translation and rotation of axes.

CONE

The equation,

$$x^2 + y^2 = a^2z^2$$

represents a circular cone whose vertex in at the origin and whose axis coincides with the z-axis. The semi-vertical angle α of the cone is given by,

$$\alpha = \tan^{-1} a.$$

In this context, the word "vertical" has nothing to do with "upright"; it merely means "of or at the vertex".

The equation,

$$\frac{x^2}{a^2} + \frac{y^3}{b^2} = z^2$$

is a cone with elliptical cross-section.

If the vertex of the cone remains at the origin, but the vertex is in some arbitrary direction (described, for example, by the direction cosines, or by spherical angles θ and ϕ) the equation can be derived by a rotation of coordinate axes.

This will introduce terms in yz, zx and xy, but it will not produce any terms in x, y or z, nor will it introduce a constant. Therefore the equation to such a cone will have only second degree terms in it.

The equation will be of the form,

$$ax^2 + by^2 + cz^2 + 2fyz + 2gzx + 2hxy = 0$$

With one proviso, the converse is usually true, namely that equation represents a cone with vertex at the origin. The one provison is that if,

$$\begin{vmatrix} a & h & g \\ h & b & f \\ g & f & c \end{vmatrix} = 0$$

equation will factorize into two linear expressions, and will represent two planes, which intersect in a line that contains the origin - and this could be regarded as a special case of a cone, with zero vertical angle. If, however, the two linear factors are identical, the two planes are coincident.

The equation to a cone of semi vertical angle α whose vertex is at the origin and whose axis has direction cosines (l, m, n) can be found as follows. The equation,

$$lx + my + nz + h = 0$$

represents a plane that is perpendicular to the axis of the cone and is at a distance h from the origin (i.e. from the vertex of the cone). Let P (x, y, z) be a point in the plane and also on the surface of the cone, at a distance r from the origin. The semi vertical angle of the cone is then given by, $\cos \alpha = h/r$.

But, $r^2 = x^2 + y^2 + z^2$ and, from equation, $h^2 = (lx + my + nz)^2$. Thus, $(x^2 + y^2 + z^2)\cos^2 \alpha = (lx + my + nz)^2$ is the required equation to the cone.

It might be thought that the equation to an inclined cone is unlikely to find much application in astronomy. Here is an application. A fireball (i.e. a bright meteor, potentially capable of depositing a meteorite) moves down through the atmosphere with speed V along a straight line trajectory with direction cosines (l, m, n) referred to some coordinate system whose xy-plane is on the surface of Earth (assumed flat). If v is the speed of sound, and $V > v$, the meteoroid will generate a conical shock front of semi vertical angle α given by

$$\sin \alpha = v/V.$$

At a time t before impact, the coordinates of the vertex will be (lVt, mVt, nVt), and the equation to the conical shock front will then be

$$[(x + lVt)^2 + (y + mVt)^2 + (z + nVt)^2] \cos^2 a$$
$$= [l(x + lVt) + m(y + mVt) + n(z + nVt)]^2$$

Part of this shock front (at time t before impact) has already reached ground level, and it intersects the ground in a conic section given by putting $z = 0$ in equation:

$$(\cos^2 a - l^2)x^2 + (\cos^2 a - m^2)y^2 - 2lmxy$$

$$-2Vt\sin^2 a\,(lx + my) - V^2t^2\sin^2 a = 0$$

and witnesses on the ground on any point on this conic section will hear the shock front at the same time.

SECOND DEGREE EQUATION

The general second degree equation in three dimensions is,

$$ax^2 + by^2 + cz + 2fyz + 2gzx + 2hxy + 2ux + 2vy + 2wz + d = 0$$

This may represent a plane or pair of planes (which, if not parallel, define a straight line), or an ellipsoid, paraboloid, hyperboloid, cylinder or cone. The equation could, if convenient, be divided through by d (or any of the other constants), and there are in reality only nine independent constants. If d is zero, the surface contains the origin. If u, v and w are all zero, and the surface is an ellipsoid, hyperbolic paraboloid or a hyperboloid, the origin is at the centre of the figure. If the figure is an elliptic paraboloid, the origin is at the vertex. If u, v, w and d are all zero, the surface is a cone with the proviso mentioned. If a, b, c, f, g, h are all zero, the surface is a plane.

Let us consider a particular example:

$$3x^2 - 4y^2 + 6z^2 + 8yz - 2zx + 4xy + 14x - 10y - 4z + 5 = 0$$

We need to do two things. First we need to rotate the coordinate axes so that they are parallel to the axes. The equation referred to the axes will have no terms in yz, zx or xy. Then we need to translate the axes so that the origin is at the centre of the figure. Mathematically, we need to find the eigenvectors of the matrix

$$\begin{vmatrix} a & h & g \\ h & b & f \\ g & f & c \end{vmatrix} = \begin{vmatrix} 3 & 2 & -1 \\ 2 & -4 & 4 \\ 1 & 4 & 6 \end{vmatrix}$$

Some readers will readily know how to do this. Others may not, and may not even be quite certain what an eigenvector is. In any case, the eigenvectors are found to be

$$\begin{pmatrix} l_{12} \\ l_{22} \\ l_{32} \end{pmatrix} = \begin{pmatrix} -0.2406405 \\ +0.9142071 \\ +0.3260635 \end{pmatrix} \qquad \begin{pmatrix} l_{13} \\ l_{23} \\ l_{33} \end{pmatrix} = \begin{pmatrix} -0.9681194 \\ -0.2501441 \\ +0.0131423 \end{pmatrix}$$

with corresponding eigenvalues 7.422 7590, 5.953 0969, 3.530 3380. The elements of the eigenvectors are the direction cosines of the present coordinate axis with respect to the figure axes. To express the equation to the surface relative to coordinate axes that are parallel to the figure axes, we replace

$$x \text{ by } l_{11}x + l_{12}y + l_{13}z$$
$$y \text{ by } l_{21}x + l_{22}y + l_{23}z$$
$$z \text{ by } l_{31}x + l_{32}y + l_{33}z$$

This will make the terms in *yz, zx* and *xy* vanish; this should be checked numerically, particularly as it is easy to rotate the axes in the wrong sense. When the substitutions are made, the equation is found to be,

$$7.4227590x^2 - 5.9530969y^2 + 3.5303380z^2 - 7.9430994x - 11.2067840y - 11.1047998z + 5 = 0.$$

Notice that there are now no terms in *yz, zx* or *xy*.

Now we need to translate the origin of coordinates to the centre of the figure (or to the vertex if it is an elliptic paraboloid). It will readily be seen that this can be done by substituting,

$$x - \alpha \text{ for } x$$
$$y - \beta \text{ for } y$$
$$z - \gamma \text{ for } z$$

where

- α = (coefficient of x)/(twice the coefficient of x^2) = – 0.535 050 336
- β = (coefficient of y)/(twice the coefficient of y^2) = + 0.941 256 643
- γ = (coefficient of z)/(twice the coefficient of z^2) = – 1.572 767 224.

The equation then becomes 7.422 7590 x^2 – 5.953 0969 y^2 + 3.530 3380 z^2 – 0.583 3816 = 0

THREE DIMENSIONAL MATRICES

Numerous matrix manipulation are possible, and the subject expands greatly if we allow the elements to be complex numbers. These six, however, are particularly useful for many applications. It might be mentioned, however, that solving simultaneous linear equations by Kramers' Rule or by inverting a matrix are very inefficient ways of solving such equations, and that is not the main purpose of acquiring the above skills.

Most, or doubtless all, of the above operations are available in many modern mathematical computer packages. This is not what I mean, however, by "understand and know how to carry out". The student should carry out at least once by hand calculator, step by step, each of the above operations, and, at each step, try to understand not only the algebraic and arithmetic steps, but also try to visualize the geometric interpretation, particularly when rotating axes and calculating eigenvectors. After doing a hand calculation, you should then write a series of short computer programmes (rather than one vast, all-encompassing matrix package) for each operation, so that when, in future, you need to do any of these things, you can instantly obtain the answer without having to go through tedious calculations.

For example, in the previous section, when I needed the eigenvectors of the matrix, I was able to generate them with a single word "EIGEN" on a computer; a considerable amount of arithmetic was actually performed by the computer. On the question of testing a matrix for orthogonality, the usual application in mechanical and geometrical problems is to test a matrix of direction cosines. The tests can not only detect mistakes, but it can locate them and even suggest what the correct element should be. Tests for orthogonality are as follows. The student should try to think of the geometric interpretation of each. The sum of the products of corresponding elements in any two rows or in any two columns is zero. Every element is equal to plus or minus its cofactor.

The determinant of the matrix is plus or minus one. A minus sign in the last two tests indicates that the two sets of axes differ in chirality. This usually does not matter, and can easily be dealt with by reversing the signs of the elements of one eigenvector and of its corresponding eigenvalue.

ANGULAR BISECTORS

A circle is said to be inscribed in a triangle if it lies completely inside the triangle and touches all three sides.

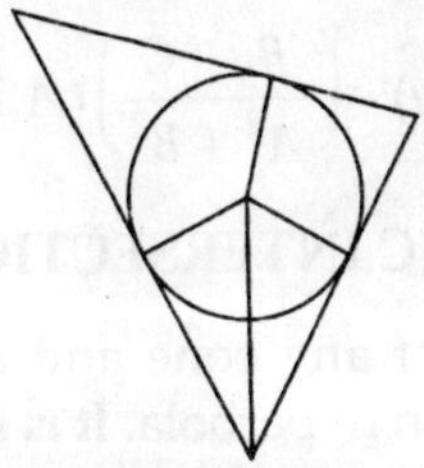

In this section we want to find:

- The centre of this circle,
- Its radius,
- The points of contact with the sides.

We will assume that the triangle is oriented so that the origin is one vertex and the positive $y$axis bisects one of its angles. Let θ be half this vertex angle. We shall also assume that the side opposite the origin is given by an equation $Ax + By = C$. The centre of the circle lies on the y-axis. Let it be $P = (0, y)$. Its distance from the left and right sides is thus guaranteed to be the same. This common distance is $y \sin \theta$. Its distance from the other side is,

$$\frac{|By - C|}{\sqrt{A^2 + B^2}}$$

so we must solve the equation,

$$y \sin \theta = \frac{|By - C|}{\sqrt{A^2 + B^2}}.$$

We do this by squaring both sides and rearranging to get,

$$y^2 \sin^2 (A^2 + B^2) = (By - C)^2$$

leading to the quadratic equation,

$$(B^2 - (A^2 + B^2) \sin^2 \theta)\, y^2 - 2BCy + C^2 = 0.$$

Exercise: The number y is one of the two roots of this equation. What is the other root? Which is the larger root? What does it mean if one root is negative?

The coordinates of the two contact points on the left and right sides are ($\pm y \sin\theta \cos\theta$, $y - y\sin^2\theta$). According to the discussion of projections, the third contact point is

$$(0,y)-\left(\frac{By-C}{A^2+B^2}\right)(A,B)$$

CONIC INTERSECTIONS

The intersection of any cone and a plane is always an ellipse, a parabola, or an hyperbola. It is simple to parametrize it, and not too difficult to tell exactly what its location and dimensions are (when the cone is rightcircular). Namely, we can essentially parametrize it as we do a circle. Recall that as t varies from 0 to 2π the point ($\cos t$, $\sin t$) runs around the unit circle in the (x, y) plane. Therefore the point,

$$(\tau \cos t, \tau \sin t, 1)$$

runs around the intersection of our cone with the plane $z = 1$, where $\tau = \tan\theta$. If the intersection is an ellipse, there will be for every point on this circle a unique point on the same line through the origin and also on the plane, of the form

$$k(\tau \cos t, \tau \sin t, 1).$$

If the plane is $Ax + By + Cz = D$ then we must have,

$$k(A\tau \cos t + B\tau \sin t + C) = D, \frac{D}{A\tau\cos t+\iota\tau\sin t+C}.$$

If the intersection is not an ellipse then this parametrization will fail when the denominator is 0. This makes sense, since a circle cannot possibly parametrize an hyperbola or parabola. It is instructive to look carefully at the denominator to see exactly when it vanishes. If

$$A\tau \cos t + B\tau \sin t + C = 0$$

this means that the point ($\tau \cos t$, $\tau \sin t$) lies on the planar line

$$Ax + By + C = 0.$$

There are three possibilities as to exactly what the intersection of a circle and a straight line can be:

- There is no intersection.
- The line is tangent to the circle, and there is one point of intersection.

- There are two points of intersection.

How do we tell which case occurs? Recall that the geometrical significance of the coefficients A, B, C is that the direction perpendicular to the line is (A, B) and the signed distance from the origin to the line is

$$\frac{-C}{\sqrt{A^2 + B^2}}.$$

The line will therefore intersect the circle of radius τ only when,

$$\frac{|C|}{\sqrt{A^2 + B^2}} \leq \tau$$

or
$$C^2 \leq \tau^2(A^2 + B^2).$$

It will intersect in exactly one point when this is an exact equality. Furthermore we can see how to find the intersection geometrically. The two points of intersection must be symmetrically placed with respect to the perpendicular direction (A, B). Multiplying (A, B) by $\tau = C$, we may as well assume that the direction vector (A, B) just touches the circle. The angle between this direction and the angle towards the intersection points satisfies

$$\cos\alpha = \frac{-C}{\sqrt{A^2 + B^2}}.$$

We can now see that if the intersection is an ellipse we parametrize it by the whole circle; if it is a parabola we parametrize it by the circle minus a single point; and if an hyperbola we parametrize each piece of the hyperbola by one of the arcs we get by removing the two points of intersection with the line $Ax + By + C = 0$.

FOCAL POINTS

Certain points called the focal points play an important role in the theory of conic sections. It was only with Kepler's discovery of his first law of planetary motion that they became of general interest, but they were in fact discovered first for purely mathematical reasons by the Greeks, and in effect allowed the Greeks to formulate the algebraic equations describing conic sections long before coordinate systems were introduced in full generality. We shall look only at ellipses. Here the fundamental fact is that one can locate two points F_1 and F_2 inside the ellipse with the following simple property:

If a line is drawn from one focus to the periphery of the ellipse and then to the other focus, the total length of the path is independent of which point on the periphery we move to.

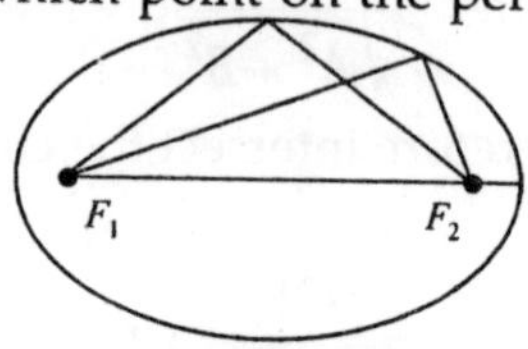

Before we begin to think about why this is true, we must locate the points $F1$ and $F2$. We can do this if we apply the characteristic property to just two paths. If a and b are the semimajor axes of the ellipse, then its equation is,

$$\left(\frac{x}{a}\right)^2 + \left(\frac{y}{b}\right)^2 = 1.$$

If $F_1 = (-f, 0)$ then $F_2 = (f, 0)$. We first consider the path from F_2 to the right hand end extremity of the ellipse, then back to F_1. Its length is $(a - f) + a + f = 2a$. On the other and we consider the path from F_2 to the top point $(0, b)$ and then to F_1. Its length must also be $2a$, and by Pythagoras' Theorem

$$f = \sqrt{a^2 - b^2}.$$

We shall see two proofs of the assertion above about the equality of path lengths. The first is algebraic. Choose a point $P = (x, y)$ on the ellipse. We can calculate the distances from F_1 to P and F_2 to P and add them them up to see whether or not we get $2a$. I leave the details as an exercise – at one point a bit of tricky algebra is required. This proof is straightforward, but not very illuminating.

5

Differential Rolling Bodies

RIGID BODIES

Rigid body motion has always been an interesting and very usable subject of classical mechanics. Many subtle points of dynamics and of mathematical techniques for studying the behaviour of physical systems can be learned by studying the motion of rigid bodies. Part of the interest of today comes from the insight that can be gained from the behaviour of spinning asteroids or artificial satellites.

Furthermore, rigid body motion can be chaotic. Besides, many everyday phenomena can be understood in terms of rigid body motion at least in a first approximation, for example balls rolling on inclines, motion in toboggans, bowls in a bowling alley, the motion of snow boarders, the dynamics of bicycles or wheels, the behaviour of billiard balls, and so on. Though the importance of rigid bodies is clear, some of the problems involving rolling particles on a surface are often modelled in beginning courses as point particles sliding on surface.

This modelling is an appropriate pedagogical device in introductory courses but we want to show here that the problems can be addressed using a rigid body approach in more advanced courses. This can be of special importance in courses attended by engineering students. In this work, we analyse the motion of a sphere that rolls without slipping on the inside of a right circular cone under the influence of a uniform gravitational field acting vertically downwards, in the direction of the symmetry axis of the cone. The motion of a spherical body rolling without slipping on surfaces of revolution has been recently studied with the purpose of illuminating control processes Here our aim is to study the motion of a sphere on the inner surface of a conical surface as an exactly solvable example of rigid body motion. We obtain the general solution of this problem expressing it in quadratures.

After reducing the problem to investigating the motion of a particle located at the sphere's centre of mass (CM) we analyse certain of its qualitative features such as the existence of a stable circular orbit For the case of a conical surface in free fall, we find that the general solution of the problem can be cast in expressions similar to those defining ellipses and hyperbolas. We also calculate the apsidal angle of the CM motion to find the orbit's symmetry axes and argue that the CM trajectory densely fills a strip on the conical surface.

EQUATIONS OF MOTION

To describe the position of the sphere's centre of mass (CM), we choose a Cartesian coordinate system (x, y, z) such that the origin of coordinates is at the position of the sphere's centre of mass (CM) when the sphere simply rests on the cone (that is, it corresponds to the vertex of the imaginary cone-shown dashed in figure as shown on which the CM moves); the coordinates are then

$$x = r \sin a \cos \vartheta, \quad y = r \sin a \sin \vartheta, \quad z = r \cos a, \qquad (1)$$

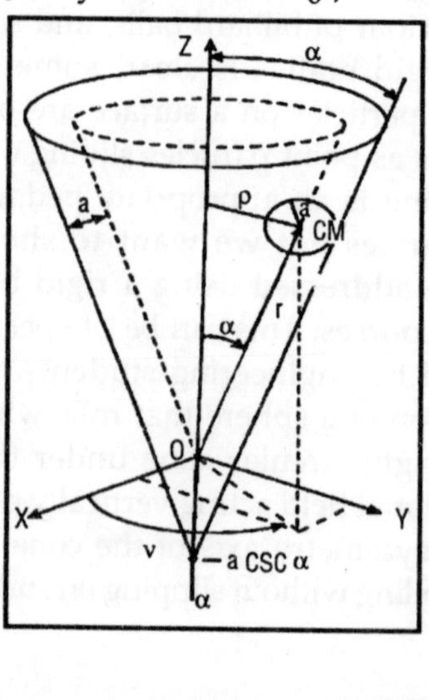

where a is half the angle of aperture of the cone-therefore, 0 _ a _ p/2, a = p/2 corresponds to motion on a plane, whereas a = 0 corresponds to motion on a cylindrical surface, r is the distance from the origin to the CM and ϑ is the polar angle of the phere's CM. All these relations may be seen from figure 1. The tangential components of the CM velocity, in the meridional and parallel directions (using a sort of geographical erminology) to the conical surface, are

$u = r', v = r \sin a \vartheta',$ (2)

respectively. Therefore, the Lagrangian of the system can be written as L = 12,

$M(u_2 + v_2) + 12$

I [×j 2 + 2×j× y cos q + × q_2 + ×j2] – Mgz, (3)

where M is the mass, I = 5Ma2/2 is the moment of inertia of the sphere, a its radius, g is the acceleration of gravity and j, q and y are Euler's angles [1]. Lagrangian (3) with u, v and z defined appropriately is fairly general; the description it gives is valid for an arbitrary surface of revolution with a sphere rolling on its surface.

The motion of the sphere is further constrained according to u – aw · ˆe1 = 0 and v + a? · ˆe2 = 0, where w is the angular velocity vector of the sphere and ˆe1 = (sin a cos J, sin a sin J, cos a), $\hat{e}_2$ = (–sin J, cos J, 0) are unit vectors in the direction of the meridional line and the parallel circle on the cone, respectively.

SPHERE TILING

Let us consider a sphere of unit radius rolling (at a constant rate, without slipping or spinning) clockwise around a circle of radius *R* on a flat plane. For what values of *R* will the sphere periodically be returned to its initial configuration (position and orientation)?

We can show that the sphere's configuration will be periodic if and only if *R* = sqrt(q^2–1) where q is a rational number greater than 1. Also, q is the ratio of the plane circle's circumference to the circumference of the circular locus of contact points on the sphere. Thus, if q = *m*/*n* the sphere will resume its initial configuration every n times around the circle.

As an example, setting q = 13/7, we find that a unit sphere rolling around a plane circle of radius 2sqrt(30)/7 = 1.56492... will return to its initial configuration every 7 times around the circle. Obviously the smallest circle that has a period of 1 (i.e., for which the sphere's configuration repeats with each trip around the circle) has a radius of sqrt(3) = 1.73205..., whereas the smallest circle with a period of 2 has a radius of sqrt(5)/2 = 1.11803... In general, the smallest circle with period n has radius sqrt($2n$+1)/n.

The equation R = sqrt(q^2–1) can be derived very;l,simply if we recognize that the pure rolling motion of a sphere in a circular path on the plane is identical to the motion of a sphere inscribed in a cone rolling on the plane, as illustrated below.

The circular locus of contact points is just the points of tangency between the sphere and the cone, and the relations between R and the circumference of that contact circle can be inferred directly from the diagram. Incidentally, this shows that for a "right" cone-which means the sphere and the plane circle have the same radius-the ratio of contact circumferences is q = sqrt($1+1^2$) = sqrt(2), so the sphere will never return precisely to its initial configuration.

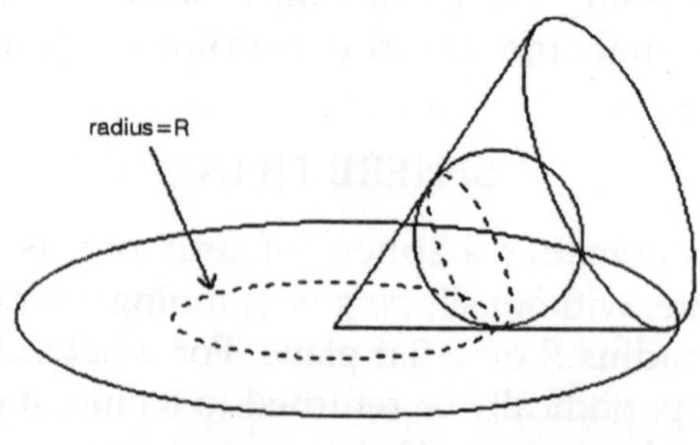

Now consider an arbitrary curve drawn on a flat surface, and a sphere rolling (without slipping or spinning) along this curve. We can parameterize the curve by giving the tangent angle "a" (relative to an arbitrarily chosen axis) as a function of the distance along the curve. If the sphere rolls along a straight line with angle "a" for a finite distance L, the change in the orientation of the sphere is given by multiplying the initial orientation by the rotation matrix,

$$M = \begin{bmatrix} ccC + ss & cs(1-C) & -cS \\ cs(1-C) & ssC + cc & sS \\ cS & -sS & C \end{bmatrix}$$

where $c = \cos(a)$, $s = \sin(a)$, $C = \cos(L/r)$, and $S = \sin(L/r)$ and r is the radius of the sphere. The net change in orientation after the sphere has rolled along k linear segments, each of length L with the path angles a_1, $a_2 ... a_k$, is given by the matrix product

$$M(\text{net}) = M(a_1)\, M(a_2) .... M(a_k)$$

As L approaches the infintesimal dL we see that S goes to dL/r and C differs from 1 by only a second-order term in (dL/r), so the infintesimal rotation matrix is,

$$m(a) = \begin{bmatrix} 1 & 0 & -\cos(a)(dL/r) \\ 0 & 1 & \sin(a)(dL/r) \\ \cos(a)(dL/r) & -\sin(a)(dL/r) & 1 \end{bmatrix}.$$

For a unit circle path (i.e., $R = 1$) we have $L = a$, and therefore $dL = da$. If we take an arc of angular size q on this path, and divide it into n equal segments, we have $da = q/n$, and for the kth segment we have $a = kq/n$.

Making these substutions and evaluating the product $m(kq/n)$ for $k = 1$ to n in the limit as n goes to infinity, we find that the net effect of rolling a sphere of radius $r = 1/\text{sqrt}(3)$ all the way around a unit circle is the identity matrix. In other words, the sphere is returned to its original orientation, in agreement with the fact that a unit sphere rolling once around a circle of radius $R = \text{sqrt}(3)$ is returned to its original orientation, as we saw previously. We would like to have a simple expression for products of matrices in the limit as the individual path lengths become vanishingly small and k goes to infinity.

To do this we might recall a similar situation with respect to products of real numbers. For example, consider how we would determine the geometric mean of the real numbers between 0 and 1. The geometric mean G of a finite set of values v_1, $v_2 .., v_k$ is simply the quantity $G = (v_1 * v_2 * ... * v_k)^{(1/k)}$, so the natural logarithm of G is,

$$\ln(G) = \frac{\ln(v1) + \ln(v2) + ... + \ln(vk)}{k}.$$

This makes it clear that the logarithm of the geometric mean of a continuous function $v(x)$ from $x = a$ to b is,

$$\ln(G) = \frac{1}{b-a} \int_{x=a}^{b} \ln(v(x))\, dx.$$

For the simple case v(x) = x with x between 0 and 1 this gives ln(G) = –1, and so the geometric mean of the real numbers from 0 to 1 is $G = 1/e$.

Now, by analogy, it might be tempting to suppose that the logarithm of M(net) in is just the sum of the logarithms of the matricies $M(a_j)$, $j = 1,2..,k$. If this were true, we could proceed to the continuous case, letting $m(a)$ denote the infintessimal rotation matrix for an incremental movement along the path, and we would have

$$\ln(M(net)) = \int_{L_0}^{L_1} \ln[m(a(L))]\, dL$$

which would imply that the net re-orientation of the sphere after rolling along the curve from L_0 to L_1 is given by the exponential of this ln(M(net)).

However, this assumes the logarithmic property log(AB) = log(A) + log(B) applies to matrices, which is not generally true, because the logarithmic property does not apply to non-commuting entities-such as general 3D rotation matricies. To see this, notice that exp(A)exp(B) does not generally equal exp(A + B), because this would imply exp(A) exp(B) = exp(B) exp(A).

In view of this, it seems that some sort of non-commuting "addition" would be needed in order to recapture the familiar logarithmic property for arbitrary matrices.

It would then be necessary to define the analagous "ordered" integration, so the integral along a path between any two points would depend on the direction in which the path was traversed.

Furthermore, the net effect of rolling once around a given closed path will generally depend on the starting point, because if we consider two points P_1 and P_2 on a closed loop, and let A denote the rotation matrix given by rolling the sphere (clockwise) along the curve from P_1 to P_2, and B the rotation matrix given by rolling on from P_2 around the rest of the loop to P_1, then the net re-orientation going clockwise around the loop from P_1 to P_1 is AB, whereas the net re-orientation going clockwise around the loop from P_2 to P_2 is BA, which need not equal AB for general paths.

Nevertheless, the above formulas are applicable to rolling along a restricted class of self-similar curves (such as straight lines), so it's interesting to develop the formulas for rolling along an arbitrary path based on the premise (which admittedly has only limited validity) that the logarithmic property applies to the rotation matrices.

We could also consider the cases where rolling along a given path length yields the same re-orientation of the sphere, and the cases where (for closed loops) the net re-orientation around the loop is independent of the starting point. For example, this condition is true for perfectly circular paths on the plane.

Notice that the infintesimal rotation matrix (3) can be split into two parts,

$$m = \begin{bmatrix} 1 & 0 & 0 \\ 0 & 1 & 0 \\ 0 & 0 & 1 \end{bmatrix} + \frac{1}{r}\begin{bmatrix} 0 & 0 & -\cos(a) \\ 0 & 0 & \sin(a) \\ \cos(a) & -\sin(a) & 0 \end{bmatrix} \mathrm{d}L.$$

This fits nicely with the power series expansion for the logarithm,

$$x^2\ x^3 \ln(a+x) = x - \frac{x^2}{2} + \frac{x^2}{3} - etc.$$

Letting "I" denote the identity matrix, and $u(a)$ denote the right hand matrix, we can write,

$$\ln(m) = \ln-\left\langle I + \frac{u(a)}{r}\mathrm{d}L \right\rangle - \frac{u(a)}{r}\mathrm{d}L - \frac{u(a)^2}{2r^2}(\mathrm{d}L) + \ldots.$$

so all terms after the first are second or higher order in dL/r. Thus, in the limit as dL -> 0 we have simply ln(m) = $u(a)/r$ dL. As a result, if L is the path-length parameter of a plane curve that makes an angle $a(L)$ relative to some arbitrary axis, then (under the assumption of the logarithmic property) the net rotation of a sphere rolling along that curve from L_0 to L_1 is given by the matrix,

$$\exp\left[\int_{L_0}^{L_1} \frac{u(a)(L))}{r} dL\right].$$

If $a(L)$ is constant we recover equation for finite rolling along a straight line. Note that these rotations are indeed commutative, so the formula is applicable. On the other hand, if we take $a(L) = L$ (yielding a circular path of radius $R = 1$) and evaluate for this case (with $L_0 = 0$ and $L_1 = q$), we arrive at the net rotation

$$\exp\begin{bmatrix} 0 & 0 & \frac{-\sin(q)}{r} \\ 0 & 0 & \frac{1-\cos(q)}{r} \\ \frac{\sin(q)}{r} & \frac{1+\cos(q)}{r} & 0 \end{bmatrix}.$$

Unfortunately, due to the inapplicability of the logarithmic property for matrices, this does not correspond to the actual rotation of a sphere rolling around a circular path, as can be seen from the fact that it gives the identity rotation whenever $q = 2pi$, regardless of the value of r, which (as we saw previously) is not correct.

RANDOM TILING

Suppose we place an equilateral triangle with edge length 1 on the surface of a sphere of radius $R >> 1$ (say, 100). Now randomly select one of the three edges of this triangle and construct an identical triangle on that edge (also on the surface of the sphere).

Now we have four exposed edges, each of which is eligible for the construction of another equilateral triangle. We choose one of these edges at random and construct another triangle. This leaves us with 5 exposed edges, all eligible, from which we randomly select one and construct another equilateral triangle, and so on.

Before long there will be some "exposed edges" on which it's not possible to construct an equilateral triangle (without overlapping an existing triangle), so we say those edges are not eligible, and we exclude them from our random selection list. At the beginning of this process our list of eligible edges will obviously increase by 1 at each step, but then it will occassionally not increase on some steps, and eventually it will start to decrease as we continue to adjoin more triangles. Ultimately we will have no more eligible edges.

What fraction of the sphere's surface area would we expect to cover in this way? What is the asymptotic behaviour of this fraction as R increases? (Since this process would completely tile the plane, it might seem that the coverage ratio limit as R approaches infinity must be 1, but perhaps there's a tendancy for each local region to "waste" a certain amount of area due to certain patterns of inter-ference that are repeated throughout the surface for ANY finite R, even though all the interference vanishes when the surface is perfectly flat.) How many ineligible exposed edges would be present at the end?

If, instead of randomly deciding where to add the next triangle, we are allowed to plan our moves, is there a simple algorithmic strategy that maximizes the covered area and/or minimizes the final number of ineligible exposed edges?

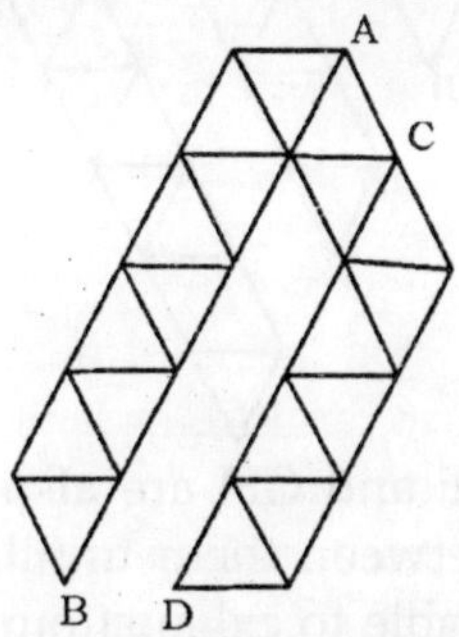

Some people have suggested that, since it's possible to fit 5 but not 6 equilateral triangles to a common vertex, the coverage must be around 5/6, regardless of whether the pattern was random or not. However, the coverage may actually be less than that, and it can depend on the sequence in which the tiles are placed. We could ask the opposite question, namely, what is the LEAST number of tiles that leave no eligible edges remaining. Suppose we have a configuration like this on the surface of a very large sphere:

The lines AB and CD are nearly geodesics and essentially parallel as they cross the segment AC. However, they don't remain parallel (as they would on a perfectly flat plane). Instead they are slowly converging, based on the fact that the centerlines of those two rows of tiles must be geodesics, and by symmetry the cross-wise row of tiles along AC must be the point of maximum separation between those geodesics. As a result, not only is it impossible to fill in the 6th edge of the upper "hexagon", it's also impossible to place a triangle anywhere between the lines AB and CD anywhere below C. If we continue this pattern, the lines AB and CD will eventually intersect, but it will take a very long time if the curvature is very slight. Also, we could repeat this pattern, creating a series of "stripes" alternating between tiled and untilable regions, like this:

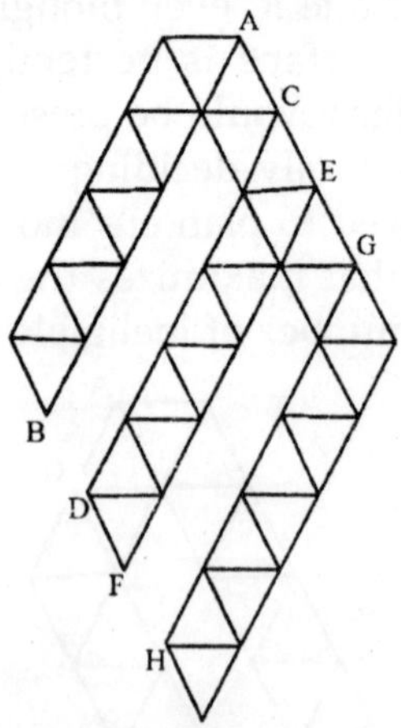

Here the lines EF and GH are also slightly converging, making the region between them unfillable. In this way we might, by design, be able to exhaust our eligible edges while having tiled only about 2/3 of the surface, since the unfillable slice between every two rows of tiles would have about half the area of a row of tiles.

Of course, this pattern requires us to carefully arrange our construction. For example, if we didn't tile all the way out to

DF on the central path, we could build a tile path from H to B, which wouldn't quite connect perfectly, but would block the central row of tiles. In general we could construct some very "maze-like" patterns, and if we just randomly selected our sites for each successive triangle, we would (or at least COULD) get some fairly jagged arrangements with extensive unfillable "slices". As a result, I'd expect to end up with less than 5/6 coverage... and for it to be possible to get significantly different coverage by design versus random selection.

It is possible to show that we can actually get arbitrarily low coverage simply by constructing one geodesic "belt" around a great cirlce of the sphere, and then constructing a triangle on each of the exposed faces of that belt. We'll end up with an arrangement that looks like this

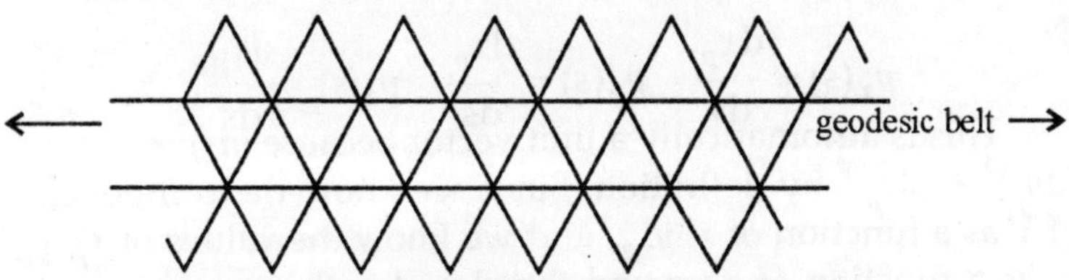

Notice that it's not possible to construct any more triangles on this configuration, since the only exposed edges are subtending and angle just slightly less than what is required to accommodate another triangle.

Thus, for an arbitrarily large sphere we have a saturated configuration of tiles that covers just an arbitrarily thin slice around a great circle. Naturally we wouldn't expect to reach this configuration by chance, but it illustrates how this type of tiling can easily become saturated even while larges regions of the sphere's surface are totally uncovered. Of course, this construction assumes the "belt" fits perfectly, but there are infinitely many values of R for which this is true.

Along these same lines, we could tile along any of the geodesic projections of the edges of a regular solid, and then "saturate" those geodesic segments by placing triangles on each exposed face, just as we did with the geodesic belt. In this way, we can get just about any amount of coverage, from the max possible all the way down to virtually zero.

Speaking of the maximum possible, we wondered if it might be possible to achieve coverages close to 1 in many cases (aside from the obvious Pythagorean cases), because it might be possible to construct a spiral belt, with "dog legs" as needed in order to circle the sphere and end up just skimming the previous loop, without leaving much empty space.

D'ALEMBERT'S PARADOX

For a vector function $V(x, y, z)$ in space, let v_x, v_y, and v_z denote the components of V. The *circulation* of the vector field V around any simple closed path S is defined as the integral of the tangential component of V around that path (in the "right-handed" direction). If the path is defined parametrically as a function of the path length parameter s by equations for $x_p(s)$, $y_p(s)$, $z_p(s)$, then the unit tangent vector as a function of s is,

$$p_x(s) = \frac{dx_p}{ds} \quad p_y(s) = \frac{dx_p}{ds} \quad p_z(s) = \frac{dx_p}{ds}.$$

This is automatically a unit vector because $(ds)^2 = (dx_p)^2 + (dy_p)^2 + (dz_p)^2$ by definition. Since we know the components of V as a function of x, y, z, and we know the values of x_p, y_p, z_p as a function of s around the closed path, we can express the components of V as functions of s on the path.

Thus we have the vector functions $V(s)$ and $p(s)$ for points on the path. Recall that the dot product of two vectors is a scalar whose magnitude equals the product of the magnitudes of the arguments times the cosine of the angle between them. Thus the dot product of V and p is the projection of V along the tangent direction, and so the circulation of V around the path S is simply the integral

$$Circ(V, S) = \int_s^0 [V(s).p(s)]\, ds.$$

If the path S is planar, we can also determine the enclosed area by means of a simple integration. For example, if the path lies entirely in the xy plane, the area is given by,

$$A(S) = \frac{1}{2}\int_s^0 \left(y\frac{dx}{ds} - x\frac{dy}{ds} \right) ds.$$

At every point in space we can define "curl *V*" as a vector whose components in each of the three principle directions equal the circulation around a closed path, in the plane normal to that direction, divided by the area enclosed by the path in the limit as the area goes to zero. Without loss of generality we can assume square paths, so the *z* component of the curl of *V* can be expressed as the limit of the circulation around a square path in a plane parallel to the *xy* plane.

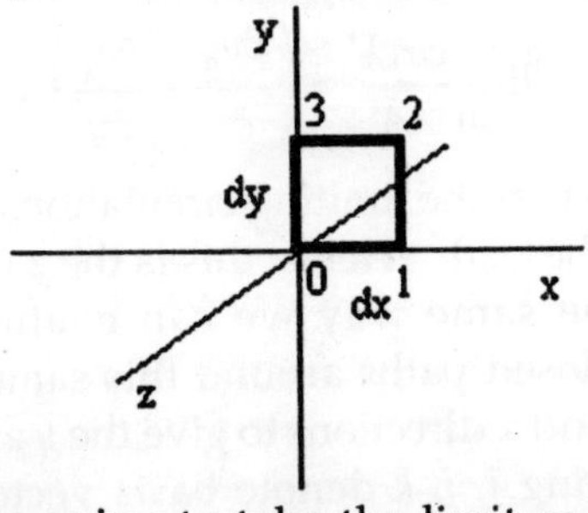

Since we are going to take the limit as the area goes to zero, and all the partial derivatives *f* the vector field are finite (by assumption), we can choose a small enough square so that the field is planar over this region.

In other words, the change in the components of *V* from 0 to 1 is the same as the change from 2 to 3, and the change from 1 to 2 is the same as from 0 to 3. Also, the components vary linearly along each edge. In addition, notice that the projection of *V* along each edge of the square is just the respective component of *V*. Therefore, the circulation around this square path is

$$circ = \left(\frac{v_{x0} + v_{x1}}{2}\right) dx + \left(\frac{v_{y1} + v_{y2}}{2}\right) dy - \left(\frac{v_{x1} + v_{x3}}{2}\right) dx - \left(\frac{v_{y3} + v_{y0}}{2}\right) dy$$

$$= \frac{1}{2}\left(\frac{v_{y1} + v_{y0}}{dx} dx + \frac{v_{x2} + v_{x3}}{dx} dx\right) dy - \frac{1}{2}\left(\frac{v_{x3} + v_{x0}}{dy} dy + \frac{v_{x2} - v_{x1}}{dy} dy\right) dx.$$

Since the changes in the components along opposite edges are equal, this reduces to,

$$circV = \left(\frac{\partial v_y}{\partial x} - \frac{\partial v_x}{\partial y}\right) dxdy.$$

Obviously the area enclosed by this path is simply $dxdy$, so the limit of the circulation in the plane normal to the z direction as the area of the path goes to zero is

$$\lim_{A\to 0} \frac{circV}{A} = \left(\frac{\partial v_y}{\partial x} - \frac{\partial v_x}{\partial y}\right).$$

(It can be shown that the limiting circulations are independent of the shape of the path.) Hence this is the z component of the curl of V. In the same way we can evaluate the limiting circulation for closed paths around this same point in planes normal to the y and x directions to give the y and x components of the curl. Letting i, j, k denote basis vectors in the x, y, z directions respectively, the curl can be written as

$$circV = \left(\frac{\partial v_z}{\partial y} - \frac{\partial v_y}{\partial z}\right) i + \left(\frac{\partial v_x}{\partial z} - \frac{\partial v_z}{\partial x}\right) j + \left(\frac{\partial v_y}{\partial x} - \frac{\partial v_x}{\partial y}\right) k.$$

In fluid mechanics if the vector field V represents the velocity of the fluid at each point, then curl V is called the *vorticity* of the fluid at that point. The vorticity is just twice the angular velocity vector. If the vorticity of a flowing fluid is everywhere zero, then the flow is said to be irrotational. Kelvin showed that vorticity is conserved in an ideal non-viscous incompressible fluid, so if the fluid is irrotational at any instant, it remains irrotational thereafter.

NON-EUCLIDEAN SPACES

Is it possible to isometrically embed a non-Euclidean manifold in a Euclidean manifold of higher dimension? If we limit ourselves to just ONE new dimension the answer is no. This was proved around 1901 by Hilbert, who showed that the original non-Euclidean space (the 2D hyperbolic plane of Lobachevski, Bolyai, et al) cannot be isometrically embedded in its entirety in 3D Euclidean space. However, it CAN be embedded in 6D Euclidean space, and even in 5D Euclidean space. Apparently the question of whether there exists a complete isometric embedding in 4D Euclidean space remains open. In any case, we can always embed a smooth metrical non-Euclidean space in a higher-dimensional Euclidean space, but it usually takes more than just one extra dimension.

Of course, if we consider semi-Riemannian manifolds it complicates the question a bit. If we're willing to allow imaginary "distances" (i.e., distances whose squares are negative), then it IS possible to embed the hyperbolic plane in 3D Euclidean space -as the surface of a sphere with imaginary radius. In fact, as early as 1766 Lambert observed that if you assumed there are at least two distinct lines through a given point that don't intersect a given line, then the area of a triangle with angles a, b, c would be $-R^2 (a + b + c - p_i)$ for some constant R. He knew that the area of a triangle on a real sphere of radius R was $R^2 (a + b + c - p_i)$, so he wrote "one could almost conclude that the new geometry would be true on a sphere of imaginary radius". It turns out that if you just plug in iR as the radius in any formula for spherical geometry you get the corresponding formula for hyperbolic geometry.

On the other hand, it isn't clear that a formally Euclidean space with imaginary distances is any more intuitive than a curved space with strictly real distances. In fact, the most unintuitive aspect of the relativistic "metric" of spacetime is not it's curvature, but the fact that it isn't really a metric at all.

A metric space, in the strict sense of the term, is a manifold that satisfies the triangle inequality, which is the property that leads to our intuitive impressions of "locality". In particular, locality is transitive in a metric space, meaning that if A is close to B, and B is close to C, then A can't be too far from C. Spacetime doesn't satisfy this condition. There exist points A,B,C such that the absolute distances AB and BC are both less than the Planck length (10^{-35} meters), and yet the distance AC is the radius of the observeable universe.

CONSTANT HEADINGS

The metric on the surface of a sphere of radius R in terms

of the latitude and longitude angles θ and ϕ is,

$$(\mathrm{ds})^2 = \mathrm{R}^2(\mathrm{d}\theta)^2 + \mathrm{R}^2 \sin(\theta)^2(\mathrm{d}\phi)^2.$$

Note that θ = 0 at the North Pole, and θ = π/2 at the equator. At any given location if we define x, y distance coordinates with x pointing east-west and y pointing north-south, then $dy = R d\theta$ and $dx = R \sin(\theta)\, d\phi$. Therefore, if we fix $dy/dx = K$ for some constant non-zero K, we have

$$\frac{d\phi}{d\theta} = \frac{1}{K\sin(\theta)}.$$

Dividing the metric by $(ds)^2$ and inserting $d\phi/ds = (d\phi/d\theta)(d\theta/ds)$ gives

$$1 = R^2\left(\frac{d\theta}{ds}\right)^2\left(1+\frac{1}{K^2}\right)$$

and so we have the equations of motion

$$\frac{d\theta}{ds} = \frac{K}{R\sqrt{1+K^2}} \qquad \frac{d\phi}{ds} = \frac{K}{R\sin(\theta)\sqrt{1+K^2}}$$

which can then be integrated to give explicit expressions for θ and ϕ as a function of the path length s. The left-hand equation gives

$$\theta(s) = \theta_0 + \frac{K}{R\sqrt{1+K^2}}s.$$

Inserting this into the right hand equation and integrating gives

$$\phi(s) = \phi_0 + \frac{1}{K}\ln\left(\frac{[1-\cos(ks+\theta_0)]\sin(\theta_0)}{[1-\cos(\theta_0)]\sin(ks+\theta_0)}\right)$$

where we have put

$$k = \frac{K}{R\sqrt{1+K^2}}.$$

This curve on the surface of a sphere is sometimes called a loxodrome, rhumb line, or spherical helix. A path beginning on the equator and maintaining a constant heading of K=0.2 is shown in the figure below.

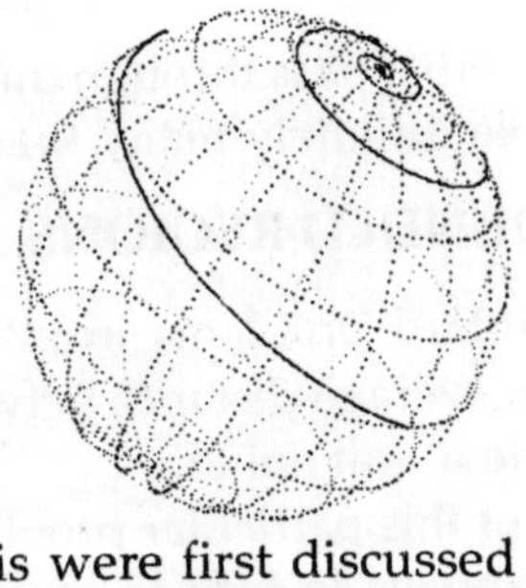

Curves like this were first discussed by Pedro Nunes in 1550. He is said to have believed that this curve represents the shortest distance between two points on the surface of a sphere, but considering the spiral nature of the curve near the poles, it's hard to imagine how anyone could have thought it was the shortest distance.

Interestingly, a constant northwest heading (for example) will reach the pole in a finite distance, but it will circle the pole infinitely many time as it approaches.

To see this, we can integrate the above equation for dq/ds for a rhumb line going from the North Pole ($\theta = 0$) to the South Pole ($\theta = \pi$). This gives the length of the complete rhumb line as

$$s = \pi R \frac{\sqrt{1+K^2}}{K}.$$

If K=0 the length becomes infinite, because that represents a purely "westward" heading. On the other hand, as K approaches infinity, the heading becomes purely "northward", and the total distance between the poles is simply half the circumference, i.e., $s = \pi R$.

To find the total longitude travelled by a rhumb line as it moves from the latitude θ_1 to the latitude θ_2, we can integrate the equation for $d\phi/d\theta$ to give

$$\phi = \frac{1}{K}\ln\left(\frac{\tan(\theta_2/2)}{\tan(\theta_1/2)}\right).$$

Obviously if either θ_1 or θ_2 equals 0 or π, the argument of the natural logarithm is either 0 or infinite, and in either case the natural log is infinite. Therefore, a rhumb line extending to either of the poles must pass through infinite longitude, i.e., it must circle the pole infinitely many times.

DISTANCE IN BOUNDED REGIONS

One of the standard problems in introductory calculus courses is to find the average distance between two randomly selected points inside a unit sphere.

The popularity of this particular problem is probably due to the fact that it happens to lead to an integral that can be evaluated in "closed form" to give a nice explicit answer, but it's interesting to consider the average distances (and powers of distances) within various other bounded regions in various dimensions.

For regions other than a sphere, the solution is not always so simple, but we'll find that a parametric formulation of the distance density can be used to simplify the analysis.

We'll begin by considering circles (spheres) and squares (cubes) of different dimensions, starting with the classic problem of the solid unit sphere in 3D space. Let R and r be the radial distances from the origin to two randomly chosen points in a unit sphere, and let θ be the angle between these vectors. The distance between the two points is

$$s = \sqrt{R^2 - 2Rr\cos(\theta) + r^2}.$$

Covering just the case $R > r$, we need to triple-integrate this quantity over r, R, and θ, and we need to weight the quantity in proportion to the fraction of the two-point state-space corresponding to each set of parameters. For given values of r and R, the angle θ defines a circle whose circumference is proportional to $\sin(\theta)$, so this is the weight for the θ integration. Similarly each value of r and R defines a sphere with surface area proportional to r^2 and R^2 respectively, so these are the weights for the r and R integrations.

We can restrict our analysis to just the case $R > r$ because the other case ($R < r$) is symmetrical and has the same distribution of distances. Hence we need only integrate the distance function with the appropriate weights for the ranges $r = [0, 1]$, $R = [r, 1]$, and $\theta = [0, \pi]$, and then divide the result by the triple integral of just the weights $(rR)^2 \sin(\theta)$, which is 1/9.

This gives the mean distance,

$$\bar{s} = 9\int_0^1 r^2 \int_r^1 R^2 \int_0^\pi \sin(\theta) = \sqrt{R^2 + r^2 - 2Rr\cos(\theta)} = \frac{36}{35}.$$

We can also evaluate this integral with the distance function raised to any integral power, to give the average of the nth powers of the distances

$$\overline{s^n} = \frac{(72)2^n}{(n+2)(n+4)(n+6)}.$$

Notice that the weight for θ is very fortuitous, because the factor of sin(θ) enables us to evaluate the integral in closed-form, using the easy integral for

$$\sin(x)\sqrt{A - B\cos(x)}.$$

In contrast, the seemingly simpler case of a unit *disk* is actually more difficult because it lacks this convenient weight factor. It's also worth noting that if we try to cover both the cases $R > r$ and $R < r$ at once by integrating over r = [0, 1] and R = [0, 1] we are led to a result like 21/20 instead of 36/35.

The subtle problem here is that the integral of,

$$\sin(\theta)\sqrt{R^2 + r^2 - 2Rr\cos(\theta)}$$

over the range θ = [0,π], divided by the integral of the weight, is,

$$\frac{\left[(R+r)^2\right]^{3/2} - \left[(R-r)^2\right]^{3/2}}{6Rr}$$

which is formally symmetrical in R and r. Now, since $R + r$ is always non-negative there's not much ambiguity in evaluating the left hand term in the numerator; we just take the positive value $(R + r)^3$, glossing over the fact that there's really a square root there in the 3/2 power (which means we *could* have taken the negative root).

However, the right hand term requires a bit of thought: Do we evaluate it as $(R - r)^3$ or $(r - R)^3$? The answer depends on whether $R > r$ or $R < r$.

In these two cases the above expression reduces to,

$$R+\frac{r^2}{3R}\ if\ R>r \qquad \mathrm{r}+\frac{R^2}{3r}\ if\ R<r$$

so it's necessary (not just convenient) to treat the cases separately. Fortunately, due to symmetry, the cases give the same distribution of distances, so we need only evaluate one of them.

Now let's consider the distances on a unit disk. This case can be formulated in essentially the same way as with the unit sphere, except that the weights are different.

Each angle θ now represents only a single point, so it's weight is just 1. Each radius r and R now represents a circle with length proportional to r and R respectively, so these are the weights. The triple integral of the product of these weights is π/8, so the average distance on a unit disk can be expressed as

$$\bar{s}=\frac{8}{\pi}\int_0^1 r\int_0^1 R\int_0^\pi \sqrt{R^2+r^2-2Rr\cos(\theta)}\ d\theta\ \mathrm{dR}\ \mathrm{dr}.$$

Unfortunately this integral isn't as easy to evaluate as the one for the sphere, because it lacks the sin(θ) factor. We will describe a much simpler approach later, so rather than presenting the elaborate integration here, suffice it to say that the result is 128/(45π). We can also evaluate the above integral for any integer power of the distance function. For odd powers of the form $2k$–1 the general result is whereas for even powers of the form $2k$ the general result is

$$\overline{\left(s^{2k}\right)}=\frac{(2k+2)!}{[(k+1)!]^2\,(k+1)(k+2)}.$$

Thus the average squared distance between two point in a unit disk is 1, and the average cubed distance is 2048/(525π). Incidentally, this last formula shows that if C_n denotes the nth Catalan number, then the average ($2k$)th power of distances on a unit disk is just $C_{k+1}/(k+1)$.

As an alternative to the preceding triple integral, we could just integrate the unit disk by scanning two points (x, y) and (X, Y) orthogonally across the disk. The weight factors are all

1 in this case, and their product integrates to the squared area of the disk, π^2, so we have the quadruple integral

$$\overline{s} = \frac{1}{\pi^2} \int_{-1}^{1} \int_{-a}^{a} \int_{-1}^{1} \int_{-b}^{b} \sqrt{(x-X)^2 + (y-Y)^2}\, dY\, dX\, dy\, dx$$

where $a = (1 - x^2)^{1/2}$ and $b = (1 - X^2)^{1/2}$. This gives the same results as the previous triple integral, but it's even more laborious to integrate. We will show later a single integral that gives the same results. As a way of introducing the simpler method for evaluating the average distances in arbitrary bounded regions, let's first consider the distribution of distances on a 1 × 1 unit square.

We could use the rectangular approach of the preceding equation and just integrate the distances between two points (x, y) and (X, Y) as each parameter ranges from 0 to 1, but the resulting quadruple integral is not very easy to evaluate.

A much more efficient approach is to notice that the distances and densities (weight factors) on a unit square are distributed according to the parametric formulas

$$s(u,v) = \sqrt{u^2 + v^2}. \qquad \delta(u,v) = 4(1-u)(1-v).$$

Therefore, we just need to integrate the product $\delta(u, v)$ $s(u, v)$ as u and v range from 0 to 1. The orthogonal way of formulating this double-integral is

$$\overline{s} = 4 \int_0^1 (1-u) \int_0^1 (u,v)\sqrt{u^2 + v^2}\ \mathrm{dv\, du}.$$

This is a big improvement over the quadruple integral, but it still is not easily evaluated in closed form. Let's try polar coordinates in this parametric space by setting $u = r\cos(\theta)$ and $v = r\sin(\theta)$ and integrating over the region $u < v$ by letting r range from 0 to $1/\cos(\theta)$ at each θ from 0 to $\pi/4$.

For any incremental slice of θ the weight at r is proportional to r, and of course the integral of r over this region $u < v$ is just 1/2, so we have the integral

$$\overline{s} = 8 \int_0^{\pi/8} \int_0^{1/\cos(\theta)} (1 - r\sin(\theta))(1 - r\cos(\theta))\mathrm{r}^2 \mathrm{d}r\mathrm{d}\theta.$$

This value of this double-integral is,

$$\frac{2+\sqrt{2}+5\ln\left(1+\sqrt{2}\right)}{15}=0.521405433...$$

By incrementing the exponent of r in the integral we can evaluate the average of the nth powers of distances on the unit square. In general the results are of the form

$$\overline{s^n}=\frac{A_n+B_n\sqrt{2}+C_n\ln\left(1+\sqrt{2}\right)}{D_n}$$

where the values of A, B, C, D are as shown below etc.

n	A	B	C	D
1	2	1	5	15
2	1	0	0	3
3	8	17	21	210
4	17	0	0	90
5	16	73	45	1008
6	29	0	0	210
7	384	3239	1155	47520
8	187	0	0	1575

Now, what about the unit cube, or the unit $4D$ hyper-cube? The nice thing about the parametric distance density equations is that they immediately generalize to higher dimensions. The parametric equations for the distance density of a d-dimensional unit cube are

$$s(x_1,x_2,....,x_d)=\sqrt{x_1^2+x_2^2+...+x_4^2}$$

$$\delta(x_1,x_2,....,x_d)=2^d\left(1-x_1\right)\left(1-x_2\right)...\left(1-x_d\right).$$

For even powers of the distance we can immediately evaluate the d-dimensional analogs of equation. We find that the average squared distance in a d-dimensional unit cube is simply d/6.

This gives a nice trivia question: In what dimensional space is the average squared distance in a unit cube equal to unity?

The average nth powers of distances in a d-dimensional unit cube are given by the following formulas for the first few even values of n:

n	average nth power of distance in unit d-cube
2	(d)/[(1)(6)]
4	$(5d^2+7d)/[(2)(90)]$
6	$(35d^3+147d^2+88d)/[(6)(1260)]$
8	$(175d^4+1470d^3+2789d^2+606d)/[(24)(9450)]$

Returning to the parametric density formulas for the distances inside a d-dimensional cube, notice that in each case the density of the vector $v = [x_1, x_2 .... x_d]$ within a bounded region *B* is proportional to the intersection of *B* with a copy of *B* shifted by the vector *v*.

More generally, given *any* finite bounded region *B* in k-dimensional space, and any k-vector *v*, let $B[\to v]$ denote the image of *B* translated by *v*.

Then the density of distances with the magnitude and direction of *v* is proportional to the content (volume) of the intersection of *B* with $B[\to v]$.

This is because *B* is the set of possible starting points for the vector *v*, and $B[\to v]$ is the corresponding set of end points. *A* particular instance of *v* (with a particular starting point in *B*) is a distance between two points of *B* if and only if its end point is also in *B*.

Thus, the intersection of $B[\to v]$ with *B* (divided by the content of *B* itself) is the proportion of points in *B* for which the vector *v* is a distance to another point in *B*. This is illustrated for a few simple shapes below.

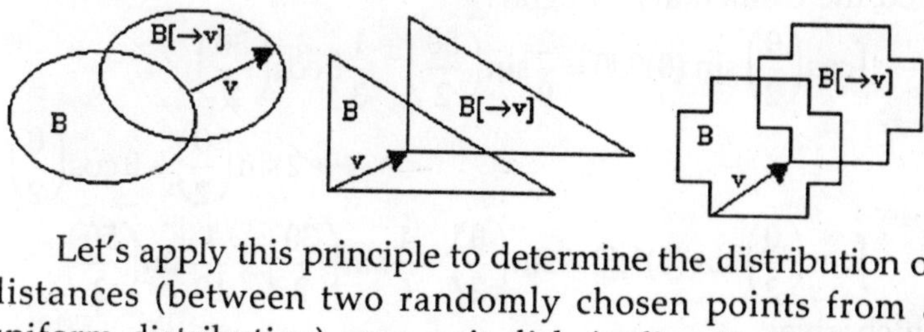

Let's apply this principle to determine the distribution of distances (between two randomly chosen points from a uniform distribution) on a unit disk (radius 1). For any 2-vector $v = [r \sin(\theta), r \cos(\theta)]$ the density is proportional to the overlap of two unit circles with centers a distance *r* apart.

This is given by,

$$4\int_{r/2}^{1}\sqrt{1-x^2}\,dx = 2inv\tan\left(\frac{\sqrt{4-r^2}}{r}\right)-\frac{r}{2}\sqrt{4-r^2}.$$

Now, in terms of the parametric lattice of distance vectors (which can be seen as the first quadrant of a circle of radius 2), the number of vectors with magnitude r is proportional to r, so the overall density of distances of magnitude r is proportional to r times the above quantity. The integral of this resulting function from $r = 0$ to 2 is simply $\pi/2$, so the distribution density of distances on a unit disk is

$$\delta(r) = \frac{r}{\pi}\left[4inv\tan\left(\frac{\sqrt{4-r^2}}{r}\right)-\sqrt{4-r^2}\right].$$

The mean value of any function f of the point-to-point distances is then given by

$$\bar{f} = \int_0^2 f(r)\delta(r)dr.$$

If we make the change of variables $r = 2\cos(\theta/2)$, the density function can be simplified and we have

$$\bar{f} = \frac{2}{\pi}\int_0^{\pi} f\left(2\cos\left(\frac{\theta}{2}\right)\right)\sin(\theta)[\theta-\sin(\theta)]\,d\theta.$$

In particular, if f is just the nth power of the distances, we replace the f function with $[2\cos(\theta/2)]^n$. Thus for $n = 1$ we just need the elementary integrals

$$\int\cos\left(\frac{\theta}{2}\right)\sin(\theta)\theta d\theta = \frac{2}{9}\sin\left(\frac{3\theta}{2}\right)-\frac{1}{3}\theta\cos\left(\frac{3\theta}{2}\right) + 2\sin\left(\frac{\theta}{2}\right)-\theta\cos\left(\frac{\theta}{2}\right)$$

$$\int\cos\left(\frac{\theta}{2}\right)\sin(\theta)^2 d\theta = \sin\left(\frac{\theta}{2}\right)-\frac{1}{6}\sin\left(\frac{3\theta}{2}\right)-\frac{1}{10}\sin\left(\frac{5\theta}{2}\right)$$

which gives,

$$\bar{s} = \frac{2}{\pi}\int_0^{\pi} 2\cos\left(\frac{\theta}{2}\right)\sin(\theta)[\theta-\sin(\theta)]\,d\theta = \frac{128}{45\pi}$$

in agreement with the mean distance computed previously by triple and quadruple integrals. An important advantage of this approach is that it not only gives a simpler integration, it gives the entire distribution explicitly, rather than just the mean. Using this same density distribution we can compute the mean value of any function of the distances. For example, the mean of the inverse distances $1/s$ in a unit disk is given by

$$\overline{\left(s^{-1}\right)} = \frac{2}{\pi}\int_0^\pi \left(2\cos\left(\frac{\theta}{2}\right)\right)^{-1} \sin(\theta)\,[\theta - \sin(\theta)]\,d\theta$$

$$= \frac{2}{\pi}\int_0^\pi \sin\left(\frac{\theta}{2}\right)[\theta - \sin(\theta)]\,d\theta = \frac{16}{3\pi}.$$

(The integral is even simpler in this case.) For a disk of radius r this is simply $16/(3\pi r)$. Thus, given n points uniformly distributed on a disk of radius r, there are $n(n-1)/2$ point-to-point distances, which implies that the expected sum of the inverse distances is $8n(n-1)/(3\pi r)$. It's interesting that while the mean reciprocal distance is finite, the mean squared reciprocal distance is not. This suggests a geometrical version of the Petersburg paradox: the game is to choose two points randomly from a uniform distribution inside a unit circle, and pay an amount proportional to the inverse square of the distance between them. Intuitively it might seem as if the payout would not be very high, but the expected value is actually infinite. The proposition that the density of v is proportional to the intersection of B with $B[\rightarrow v]$ is very general, and applies to non-convex (and even non-connected regions), not just simple regions like spheres and cubes. For example, it can be used to determine the distance distribution for points of a torus.

TORUS GRAVITY

In which direction would the net force of gravity point at various locations near a ring of matter? People sometimes think that, perhaps for reasons of symmetry, an object in the interior of a ring of matter would be drawn towards the centre, but this is not the case – at least not for objects in the plane of the ring. To see why, consider a very thin ring of mass treated as a circle of radius R in the plane, and a particle inside this ring at a distance r from the centre. Construct an arbitrary line passing through this particle, striking the ring in two opposite directions at distances L_1 and L_2. If we rotate this line about the particle through an incremental angle $d\theta$, it will sweep out sections of the ring proportional to $L_1\cos(\alpha)d\theta$ and $L_2 \cos(\alpha)d\theta$, where α is the angle the chord makes with the normals to the circle at the points of intersection. The net gravitational force exerted by these two opposing sections of the ring is proportional to the masses in these small sections divided by the squares of the distances, i.e., the force is proportional to $d\theta$ $\cos(\alpha)$ $(1/L_1 - 1/L_2)$ in the direction of the L_1 intersection point.

Hence the net force is in the direction of the closest point on the ring, directly *away* from the centre. Of course if the particle was inside a spherical *shell* instead of a circular ring, the conical region surrounding each ray through the particle would sweep out a section of the shell proportional to the *square* of the distances L_1 and L_2, which would then cancel out with the inverse squares in the force expressions, leaving a net zero force on the particle. However, in the case of a ring of matter, the distances L_1 and L_2 appear only to the first power in the mass terms, so they cancel only one of the powers in the inverse squares. This is why the net force along every ray is positive in the direction of the closer point on the ring, which implies that a particle in the interior plane of the ring would move outward towards the ring. There would be an equilibrium point at the centre of the ring, but this point would not be stable, i.e., it would be a local maximum of the potential energy.

To quantitatively determine the Newtonian gravitational force for a massive ring of radius R, we can note first that the field has cylindrical symmetry, so all that matters is the radial distance (from the ring's centre) of a test particle and its elevation above the plane of the ring. It's convenient to use cylindrical coordinates, r, θ, z, where r is the radial distance from the centre of the ring, θ is the angle in the plane of the ring, and z is the elevation above (or below) the plane of the ring. Without loss of generality we can focus on a test particle at a distance r from the origin and at an angle $\theta = 0$.

Letting ρ denote the mass per unit length along the circumference of the ring, each increment of the ring contributes the incremental mass $dm = \rho R d\theta$. This mass is at a distance s from the test particle, given by

$$\begin{aligned} s &= \sqrt{[R\cos(\theta) - r]^2 + [R\sin(\theta)]^2 + z^2} \\ &= \sqrt{R^2 - 2Rr\cos(\theta) + r^2 + z^2}. \end{aligned}$$

The corresponding incremental potential is $-dm/s$, so the total gravitational potential is given by the integral,

$$\Psi(r,z) = \int_0^{2\pi} \frac{-\rho R\, d\theta}{\sqrt{R^2 - 2Rr\cos(\theta) + r^2 + z^2}}.$$

The gravitational force in the r and z directions is given by the derivatives $d\psi/dr$ and $d\psi/dz$. There is no closed-form expression for the above integral, but we can bring the differentiations inside the integrals, giving the following expressions for the radial and axial forces,

$$F_r = -\frac{\partial \Psi}{\partial r} = \rho R \int_0^{2\pi} \frac{R\cos(\theta) - r}{\sqrt{\left(R^2 - 2Rr\cos(\theta) + r^2 + z^2\right)^{3/2}}} d\theta$$

$$F_z = -\frac{\partial \Psi}{\partial z} = \rho R \int_0^{2\pi} \frac{-z}{\sqrt{\left(R^2 - 2Rr\cos(\theta) + r^2 + z^2\right)^{3/2}}} d\theta$$

Using these equations, with $R = \rho = 1$, the radial gravitational force near a ring of matter at various elevations above the plane of the ring is plotted in the figure below.

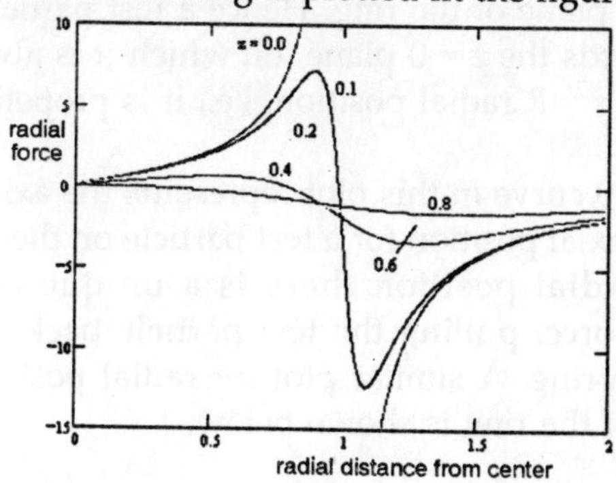

The curve labeled "z = 0.0" is for points in the plane of the ring. As discussed previously, for points away from the centre the net gravitational force is positive, i.e., outward, directly away from the centre.

However, at an elevation of 0.8R above the plane of the ring, the net radial force is negative as we move away from the central axis, meaning the radial force tends to push a test particle back towards the central axis.

Even for elevations of just $0.1R$, there is a point inside the ring's radius where the radial force passes through zero and then becomes negative. For each elevation (other than zero) there is a radial position of stable equilibrium. Of course, these locations are not equilibrium positions in the axial direction, as shown by the plot below, which gives the axial force as a function of axial distance (from the plane of the ring) for various radial positions.

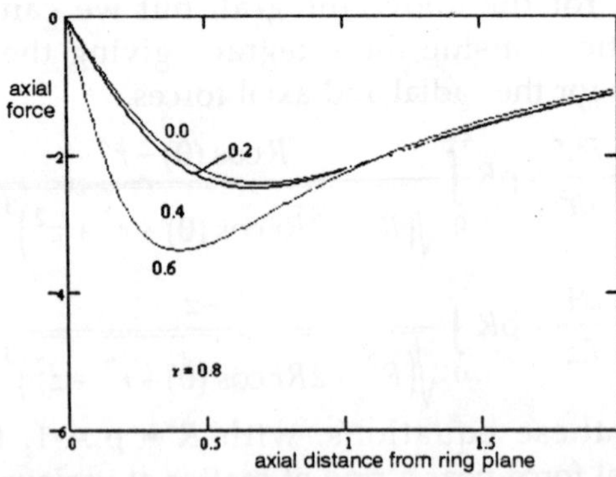

This plot shows that there is always a net axial force towards the plane of the ring. Hence a test particle is always driven towards the z = 0 plane, on which it is always driven towards the $r = R$ radial position, i.e., it is propelled towards the ring.

The $r = 0$ curve in this plot represents the axial force as a function of axial position for a test particle on the central axis. For each radial position there is a unique elevation of maximum force, pulling the test particle back towards the plane of the ring. A similar plot for radial positions outside the radius of the ring is shown below.

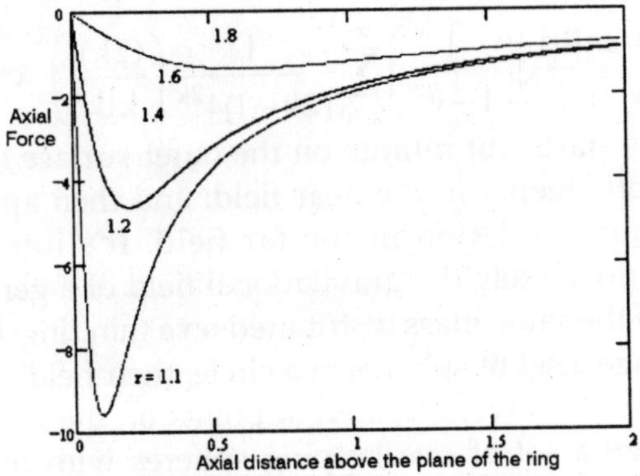

We can also express the gravitational forces as power series. If r is less than R it's convenient to write the radial force equation (with $z = 0$) in the form

$$F_r = -\frac{\rho}{R}\int_0^{2\pi}\frac{\cos(\theta)-\mu}{\left(1-2\mu\cos(\theta)+\mu^2\right)^{3/2}}d\theta$$

where $\mu = r/R$. Expanding this in powers of μ and dividing out the geometric series gives,

$$F_r = \left(\frac{\pi r\rho}{R}\right)\left(\frac{1}{1-\mu^2}\right)\left[1+\frac{1}{8}\mu^2+\frac{3}{64}\mu^4+\frac{25}{1024}\mu^6+\frac{1323}{131072}\mu^8+\ldots\right]$$

$$= \left(\frac{\pi r\rho}{R}\right)\left(\frac{1}{1-\mu^2}\right)\sum_{k=0}^{\infty}\frac{1}{(k+1)4^{2k}}\left(\frac{(2k)!}{k!k!}\right)^2\mu^{2k}.$$

Thus at $r = 0$ the radial force is zero, but at any non-zero value of r there will be a net outward force, initially proportional to r, and then increasing sharply as r approaches R. (The ring is assumed to have infinite density so the outward force goes to infinity at $r = R$.)

On the other hand, if r is greater than R, it's convenient to write the radial force equation (again with $z = 0$) in the form,

$$F_r - \frac{\rho\nu}{r}\int_0^{2\pi}\frac{\nu\cos(\theta)-1}{(\nu^2-2\nu\cos(\theta)+1)^{3/2}}d\theta$$

where $\nu = R/r$. Expanding this gives the series

$$F_r = -\left(\frac{2\pi R\rho}{r^2}\right)\left(\frac{1}{1-\nu^2}\right)\left[1-\frac{1}{4}\nu^2-\frac{3}{64}\nu^2-\frac{5}{256}\nu^8+\ldots\right]$$

$$= \left(\frac{2\pi R\rho}{r^2}\right)\left(\frac{1}{1-\nu^2}\right)\sum_{k=0}^{\infty}\frac{1}{(2k-1)4^{2k}}\left(\frac{(2k)!}{k!k!}\right)^2 \nu^{2k}.$$

So this starts out infinite on the outer surface (where ν = 1), drops off steeply in the near field, and then approach an inverse-square relation in the far field. It's interesting to consider how closely the gravitational field of a general torus is to that of the same mass distributed on a thin ring. Obviously if we arrange a set of spheres in a circle, their field is the same as if their masses were located at points on the ring. We can also arrange a set of overlapping spheres with most of the overlap removed by internal spheres, all of which are centered on the ring, so in this way we can approximate very closely a general torus with just positive and negative spheres that can be treated as points on the central ring.

QUANTUM MECHANICS

Suppose one of the observable attributes X of a physical system has the form of a binary variable x which can be either true or false, and suppose we have a physical law that tells us how the value of x changes with time in specified circumstances.

Letting a numerical value of $x = +1$ signify "true" and a numerical value of $x = -1$ signify "false", the equation(s) expressing our physical law may be of the form $x = f(t)$ where the function f takes on only the values –1 or +1. In this case we say the physical law ranges exactly over the observable states of the system.

However, it's conceivable that a physical law might range over a larger set than the set of observable states. For example, the function $f(t)$ representing the physical law governing the behaviour of x might range over the entire set of real numbers on the interval from –1 to +1, even though the only observable states correspond to the two specific values –1 and +1, as depicted below.

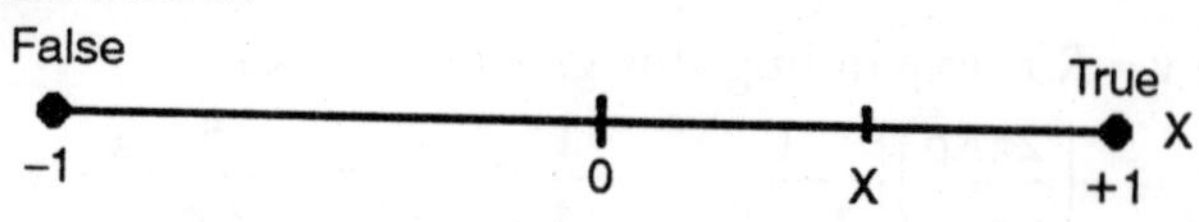

In order for the physical law to be meaningful we need some interpretation of the intermediate values of x, values which do not correspond to observable states. One possible interpretation is that the numerical value of x indicates the propensity of the system to be observed in State +1 if a measurement of the X attribute is made. If the numerical value of x is –1, 0, or +1, the probability that X will be observed in State +1 should be 0, 1/2, and 1 respectively. Naturally the probability of being observed in State –1 is the complement.

This leads us to define,

$$P\{x=+1\}=\frac{1+x}{2} \qquad P\{x=-1\}=\frac{1-x}{2}.$$

If we find that repeating the same measurement twice in a row invariably yields the same value each time, we may conclude that the state of the system "jumps" from the intermediate value to the observed value when a measurement is made. (Needless to say, this raises a number of questions about what constitutes a measurement.) We define the *uncertainty* in the attribute X as zero if x has the value –1 or +1, and as infinite if x has the value 0. The simplest function giving these values is,

$$U(X)=\frac{1-x^2}{x^2}.$$

It's convenient to represent these ideas by allowing the state of the system to be not just the points between +1 and –1 on the X axis, but at any of the points on the unit circle in the x, y plane as shown below.

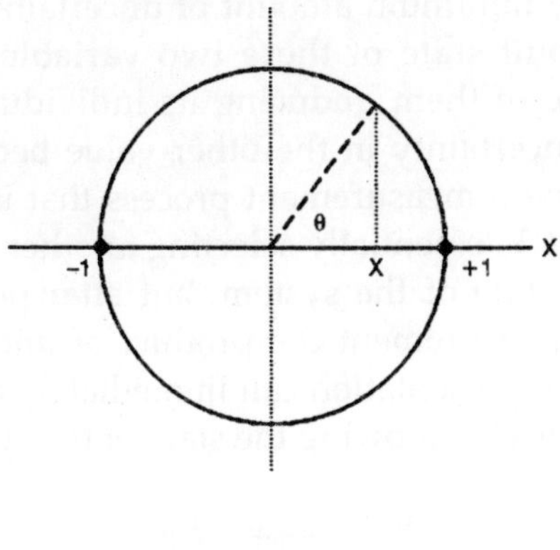

In place of the state variable x we can use the angle θ defined by $x = \cos(\theta)$. Then the probability of observing the system in State +1 is $[1 + \cos(\theta)]/2$ and the uncertainty in X is $U_X(\theta) = \tan(\theta)^2$. In this representation each variable's real axis is accompanied by a perpendicular axis, which might be regarded as the axis of another observable variable, which we will call Y, as shown below.

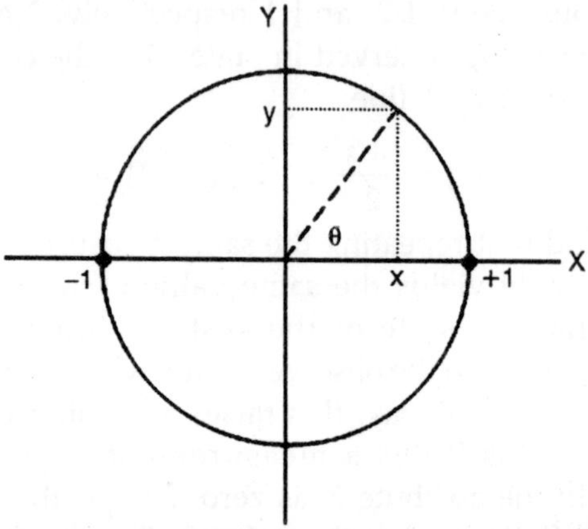

Just as with the X variable, we can also make a measurement of the Y variable, with the result that the system jumps to +1 on the Y axis (if Y is observed to be true) or −1 on the Y axis (if Y is observed to be false). In terms of the parameter θ the probability of State +1 is $[1+\sin(\theta)]/2$ and the uncertainty in Y is $U_Y(\theta) = \tan(\theta + \pi/2)^2$. The product of the uncertainties in the X and Y variables is therefore

$$U_X(\theta)U_Y(\theta) = \tan(\theta)^2 \tan(\theta + \pi/2)^2 = 1.$$

This is the minimum amount of uncertainty that we can have in the joint state of these two variables, because by measuring one of them, reducing its individual uncertainty to zero, the uncertainty in the other value becomes infinite. We can imagine a measurement process that is intermediate between X and Y, essentially selecting another axis on which to project the state of the system, but after performing this intermediate measurement the product of uncertainties will still be 1. This representation can immediately be generalized to three variables by allowing the state of the system to reside at any point on a unit sphere in three dimensions. Then depending on which measurement we perform, the system will be projected onto one of the three axes. The equations of motion (i.e., the physical laws) governing the system range over the entire surface of the sphere, but the only observable states are the six points [±1, 0, 0], [0, ±1, 0], and [0, 0, ±1]. These are the eigenvalues for rotations about the x, y, and z axes respectively.

The two non-classical features of the model described above are:

1. The laws of motion range over a larger set of states than the set of observable states,
2. Some sets of observable variables are inherently interdependent, in the sense that they can be represented by a smaller number of state variables.

6

Differential Theorems

EQUATION OF CONTINUITY

A function f is continuous at a number a if ,

$$\lim_{x \to a} f(x) = f(a)$$

The definition given above actually contains three assertions within it, neatly packaged into one limit statement.

Three requirements must be met if a function f is to be continous at a:

1. $f(a)$ must be defined (in other words, a must be in the domain of f).
2. $\lim_{x \to a} f(x)$ must exist.
3. $x \to a$.

If one of the three conditions fails at a point a, then we say that f is discontinuous at a.

DISCONTINUITIES

In this course we will encounter three basic ways in which a function may fail to be continuous.

REMOVABLE DISCONTINUITY

Consider the function,

$$f(x) = \frac{x^2 - 4}{x - 2}.$$

One may immediately observe first that a = 2 is not in the domain of f and second that after factoring the numerator and cancelling common factors, the function may be re-expressed as $f(x) = x + 2$ for x not equal to 2. The graph of $f(x)$ looks like the graph of the line $y = x + 2$ with the point (2,4) removed. We say that f has a removable discontinuity at $a = 2$ since it is possible to fill in the missing point (2,4) on the graph of f so as to obtain a new function which is continuous at f. By filling in the missing point (the gap in the graph), we remove the discontinuity.

INFINITE DISCONTINUITY

Let us now consider the function,

$$f(x) = \frac{x^2 - 4}{x - 2}.$$

Notice that $a = -1$ is the only real number not in the domain of f. Hence g is continuous everywhere except at a = –1. Consider the graph of the function: > plot((x-3)^2/(x+1)^2,x=-3..1,y=-5..50,discont= true);

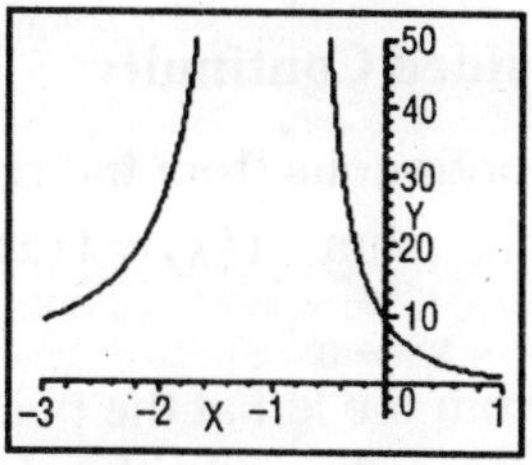

We see that,

$$\lim_{x \to (-1)} g(x) = \infty$$

In this case, g is discontinuous at a = –1 by virtue of the fact that g increases without bound as x approaches -1. We say that g has an, $a = -1$.

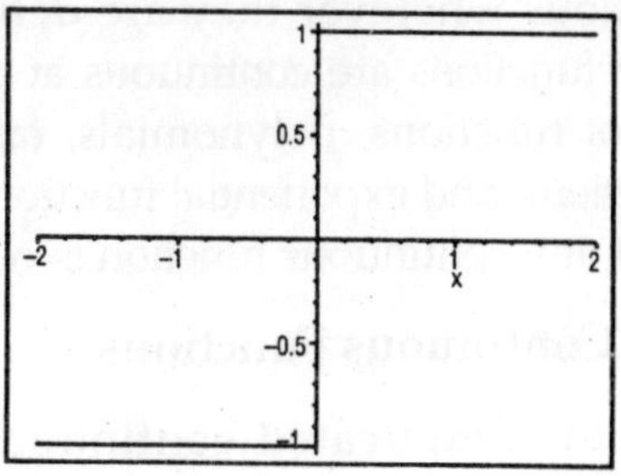

Here it is noticed that as x moves across 0 from left to right, $h(x)$ jumps from the value -1 to 1. That is, h abruptly changes as x moves from negative values to positive values. Further, h(0) is undefined and,

$$\lim_{x \to 0} h(x)$$

does not exist, so h fails to be continuous at 0. We say that h has a jump discontinuity at 0.

CONTINUITY ON AN INTERVAL

A function is continuous on an interval I if it is continuous at every point in I. Notice that we must be careful here when dealing with intervals containing endpoints where the function is not defined outside the interval. In these cases, we must introduce the ideas of continuity from the right and continuity from the left.

Definition of One-sided Continuity

A function f is continuous from the right at the point a if,

$$\lim_{x \to a^+} f(x) = f(a)$$

and f is continuous from the left at the point a if $x \to a-$. If a function is defined on only one side of an endpoint of an interval, we understand continuous at the endpoint to mean continuous from the right or continuous from the left depending on the given situation.

Common Continuous Functions

Most of the elementary functions which have been introduced so far are continuous wherever they are defined. Hence the following types of functions are continuous at every number in their domains: root functions, polynomials, rational functions, trigonometric functions, and exponential functions. Moreover, the inverse function of any continuous function is continuous.

Composition of Continuous Functions

One can build complicated continuous functions by

composing elementary continuous functions with one another by the following theorem.

Theorem: (The Composition of Continuous Functions is Continuous) If g is continuous at a and f is continuous at $g(a)$, then the composite function given by $f(g(x))$ is continuous at a. One of the basic properties of continuous functions is the subject of the following theorem.

The Intermediate Value Theorem (IVT)

If f is continuous on the closed interval $[a, b]$ and N is any number between f(a) and f(b), then there exists a number c between a and b such that $f(c) = N$.

Example: We show that $x^3 + 2x + 2 = 0$ has a real root between -1 and 0. Recall that a root of an equation is a value of the variable which satisfies the equation.

Let,

$$f(x) = x^3 + 2x + 2 \cdot$$

To show that our given equation has a real root between -1 and 0, we must show that there is a real number c such that f(c) = 0. We do this by appropriately applying the Intermediate Value Theorem.

We note that f(x), being a polynomial, is continuous everywhere. In particular, it is continuous on the closed interval [-1,0]. Evaluating f at the endpoints of this interval, we see that $f(-1) = -1$ and f(0) = 2. Hence the number $N = 0$ is between $f(-1)$ and $f(0)$. The conclusion of the Intermediate Value Theorem is that there must be a number c between -1 and 0 such that $f(c) = 0$. We have shown that there is a real root between -1 and 0. Maple allows us to verify what we have shown using the IVT.

We plot the graph of,

$$f(x) = x^3 + 2x + 2$$

and verify that the graph does indeed cross the x-axis at some number between -1 and 0.

```
> f:=x->x^3+2x+2;
```

$$f = x \to x^3 + 2x + 2$$

```
> plot(f(x),x=-1..0);
```

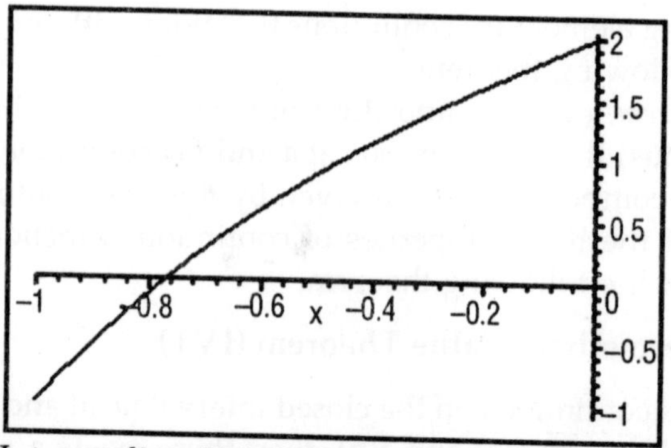

The Intermediate Value Theorem is a powerful theoretical tool in the calculus and one of the first nontrivial results we have seen so far. As an existence theorem, it does not give us specific information about the number *c*, but as we will see, knowledge of existence is enough for some interesting results later on.

INFINITY DISCONTINUITY

Often when considering piecewise defined functions, there may be "jumps" in their graphs. Consider the graph of the function,

$$h(x) = \frac{|x|}{x},$$

for example. > plot(abs(x)/x,x=-2..2,discont=true);

CONTINUITY PRINCIPLE

We will now take up the Axioms of Continuity. First, we will prove the two Continuity Principles and then discuss some of the different uses of the Continuity Axioms in our work. First, we shall need the famous *Triangle Inequality*. It is usually proved after we have given a measure to line segments, but that is not necessary.

INEQUALITY OF TRIANGLE

Proposition

If *A*, *B*, and *C* are three noncollinear points, then $AC < AB + BC$, where the sum is segment addition.

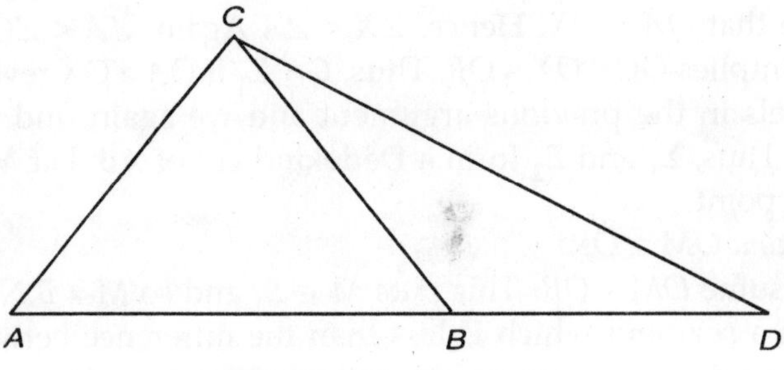

Fig. Triangle Inequality

Proof: There is a unique point D such that $A * B * D$ and $BD \cong BC$ by the first congruence axiom. Then $\angle BCD \cong \angle BDC$ by Proposition.

Now, $AD \cong AB + BD$ and since $BD \cong BC$ we have that $AD \cong AB + BC$ By Proposition $\overrightarrow{CB}$ is between the rays $\overrightarrow{CA}$ and $\overrightarrow{CD}$ Then, by definition, $\angle ACD \cong \angle BCD$ By Proposition and what we have just shown, $\angle ACD > \angle ADC$ Thus, by Proposition $AD > AC$, and we are done.

PRINCIPLE

Theorem: If one endpoint of a segment is inside a circle and the other outside, then the segment intersects the circle.

Proof: Let Γ be a circle centered at O with radius OR, so that $R \in \Gamma$ Let AB be the above segment. Let Int $(\Gamma) = \{A \mid OA < OR\}$ and Ext $(\Gamma) = \{B \mid OB > OR\}$. Let

$$\Sigma_1 = \{X \mid X \in AB \text{ and } OX < OR\}$$
$$\Sigma_2 = \{X \mid X \hat{I}\, AB \text{ and } OX \geq OR\}.$$

Note that neither of Σ_1 or Σ_2 is empty. We wish to show that these two sets form a Dedekind cut of the segment AB. There are then several things to show.

- If $Y \in AB$ then by Proposition we must have that $OX < OR$, $OX \cong OR$, or $OX > OR$ This puts X in one of the two sets. Therefore, $\Sigma_1 \cup \Sigma_2 = AB$
- Clearly, $\Sigma_1 \cap \Sigma_2 = \theta$
- We have that and
- Let and We need to show then that If then is exterior to so Thus, by Proposition$OC<OX$. Since $OX<OR$ by Proposition$OC<OR$ and

Now, assume that $OA \not\equiv OX$ Then, for the first case, assume that $OA < OX$. Hence, $\angle X < \angle A$ Again, $\angle A < \angle OCX$ which implies $OC < OX < OR$. Thus, $C \in \Sigma_1$ If $OA > OX$ reverse the labels in the previous argument and we again find that $C \in \Sigma_{1.}$ Thus, Σ_1 and Σ_2 form a Dedekind cut of AB. Let M be the cut point.

Claim: $OM \cong OR$

Assume $OM < OR$. This puts $M \in \Sigma_1$ and so $M \neq B$ Now, there is a segment which is less than the difference between OM and OR. Let M' be a point outside Γ and on AB so that MM' is congruent to that segment. Thus, we have that $M' \in \Sigma_2$ By the Triangle Inequality we have that

$$OM' < OM + MM'$$

$$OM + MM' < OR$$

$$OM' < OR$$

But, $M' \in \Sigma_2$ so that $OM' \geq OR$: a contradiction. Assuming that $OM > OR$ we use the same technique to arrive at a contradiction. Thus, $OM \cong OR$ and $M \in \Gamma$.

ANGLES AND SEGMENTS

To avoid some of the difficulties that we faced in the previous two proofs, and to facilitate matters at a later time, we will introduce a measure for angles and for segments.

The proof of the Theorem requires the axioms of continuity for the first time. The axioms of continuity are not needed if one merely wants to define the addition for congruence classes of segments and then prove the triangle inequality for these congruence classes.

It is in order to prove several of our theorems that we need the measurement of angles and segments by real numbers, and for such measurement Archimedes's axiom is required. However, the fourth and eleventh parts of Theorem, the proofs for which require Dedekind's axiom, are never used in proofs in the text.

It is possible to introduce coordinates without the continuity axioms, as in discussed in Appendix B of the text. The notation $\angle A^\circ$ will be used for the number of degrees in $\angle A$, and the length of segment AB will be denoted by $\overline{AB}$.

Theorem: There is a unique way of assigning a degree measure to each angle such that the following properties hold:

- Đ*A*° is a real number such that 0 < Đ*A*° < 180°.
- Đ*A*° = 90° if and only if Đ*A* is a right angle.
- Đ*A*° = Đ*B*° if and only if Đ*A* ≅ Đ*B*
- *If* $\overrightarrow{AC}$ is interior to *P* Ï *l* then Đ*DAB*° = Đ*DAC*° + Đ*CAB*°.
- For every real number *x* between 0 and 180, there exists an angle Đ*A* such that Đ*A*° = *x*°
- *If* Đ*B* is supplementary to Đ*A* then Đ*A*° + Đ*B*° = 180°.
- *If* Đ*A*° > Đ*B*° and only if Đ*A* > Đ*B*.

Given a segment OI, called the unit segment. Then there is a unique way of assigning a length $\overline{AB}$ to each segment AB such that the following properties hold:

- $\overline{AB}$ is a positive real number and $\overline{OI}$ = 1.
- $\overline{AB} = \overline{CD}$ if and only if *AB* ≅ *CD*.
- *A* * *B* * C if and only if $\overline{AC} = \overline{AB} + \overline{BC}$.
- $\overline{AB} < \overline{CD}$ if and only if *AB* < *CD*.
- For every positive real number *x*, there exists a segment *AB* such that $\overline{AB} = x$.

Definition: An angle $\angle A$ is acuteif $\angle A° < 90°$ and is obtuseif $\angle A° > 90°$.

Corollary: The sum of the degree measures of any two angles of a triangle is less than 180°. This follows from the Exterior Angle Theorem.

Proof: We want to show that $\angle A° + \angle B° < 180°$ From the Exterior Angle Theorem,

$$\angle A° < \angle CBD°$$

$$\angle A° + \angle B° < \angle CBD° + \angle B° = 180°,$$

Since they are supplementary angles.

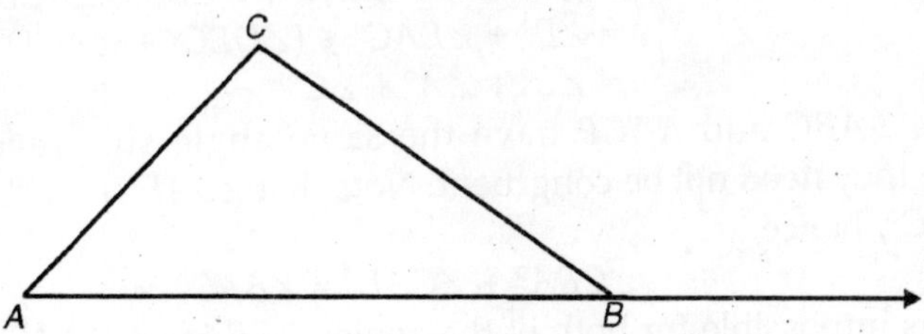

Triangle Inequality

Corollary: If A, B, and C are three noncollinear points, then $\overline{AC}$

$< \overline{AB} + \overline{BC}$. Theoremoffers an easier proof of this than the one that we gave.

SACCHERI-LEGENDRE THEOREM

This theorem gives us a setting for our later exploration into noneuclidean geometry.

STATEMENT

Theorem: The sum of the degree measures of the three angles in any triangle is less than or equal to 180°;

$$\angle A^\circ + \angle B^\circ + \angle C^\circ \leq 180^\circ.$$

Proof: Let us assume not; *i.e.,* assume that we have a triangle ΔABC in which ÐA° + ÐB° + ÐC° > 180° So there is an $x \in \mathbb{R}^+$ so that

$$\angle A^\circ + \angle B^\circ + \angle C^\circ = 180^\circ + x.$$

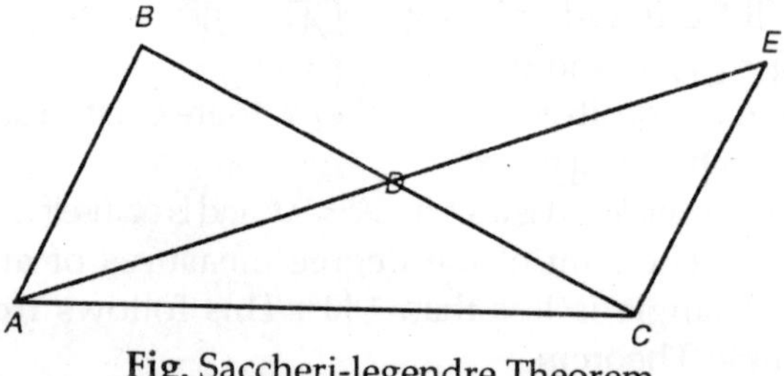

Fig. Saccheri-legendre Theorem

Let D be the midpoint of BC and let E be the unique point on $\overrightarrow{AD}$ so that $DE \cong AD$ Then by SAS $\Delta BAD \cong \Delta CED$. This makes, $\angle B^\circ = \angle DCE^\circ$ $\angle E^\circ = \angle BAD^\circ$.

Thus,

$$\begin{aligned}\angle A^\circ + \angle B^\circ + \angle C^\circ &= (\angle BAD^\circ + \angle EAC^\circ) + \angle B^\circ + \angle ACB^\circ \\ &= \angle E^\circ + \angle EAC^\circ + (\angle DEC^\circ + \angle ACD^\circ) \\ &= \angle E^\circ + \angle A^\circ + \angle C^\circ\end{aligned}$$

So, ΔABC and ΔACE have the same angle sum, even though they need not be congruent. Note that $\angle BAE^\circ + \angle CAE^\circ = \angle BAC^\circ$, hence,

$$\angle CEA^\circ + \angle CAE^\circ = \angle BAC^\circ$$

It is impossible for both of the angles $\angle CEA$and $\angle CAE$ to have angle measure greater than 1/2 $\angle BAC^\circ$, so at least one of the angles has angle measure greater than or equal to 1/2 $\angle BAC^\circ$.

Therefore, there is a triangle ΔACE so that the angle sum is$180° + x$ but in which one angle has measure less than or equal to 1/2 $\angle A°$ Repeat this construction to get another triangle with angle sum $180° + x$ but in which one angle has measure less than or equal to 1/4 $\angle A°$ Now there is an $n \in \mathbb{Z}^+$ so that,

$$\frac{1}{2^n} \angle A° \leq x,$$

By the Archimedean property of the real numbers. Thus, after a finite number of iterations of the above construction we obtain a triangle with angle sum $180° + x$. in which one angle has measure less than or equal to x Then the other two angles must sum to a number greater than 180°.

Corollary: In ΔABC the sum of the degree measures of two angles is less than or equal to the degree measure of their remote exterior angle.

INTERMEDIATE VALUE THEOREM

The Intermediate Value Theorem says that if a continuous function takes on two values y_1 and y_2 at points a and b, it also takes on every value between y1 and y_2 at some point between a and b.

A typical argument using the IVT is:

Know: $e^0 = 1$ and $e^1 = e$, $1 < 2 < e$, e^x is continuous for all real x.

Conclude: $e^x = 2$ for some x with $0 \leq x \leq 1$.

Graphically The IVT says that if y1 = f(a) and y^2 = f(b) for a function f(x), and if we draw the horizontal line $y = y_0$ for any y_0 between y_1 and y_2, then the horizontal line intersects the graph of y = f(x) at (at least one) point whose x-coordinate is between a and b.

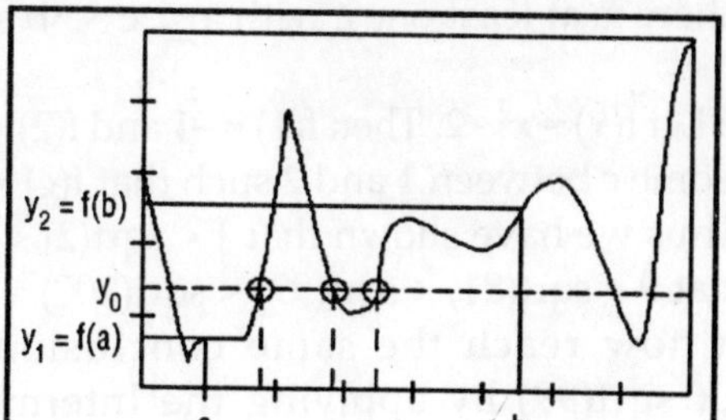

DESCRIPTION

Let f(x) be a function which is continuous on the closed interval [a,b] and let y_0 be a real number lying between f(a) and f(b), i.e. with $f(a) \leq y_0 \leq f(b)$ or $f(b) \leq y_0 \leq f(a)$. Then there is at least one c with $a \leq c \leq b$ such that $y_0 = f(c)$.

Note: "continuous on the closed interval [a.b]" means that f(x) is continuous at every point x with a < x < b and that f(x) is right-continuous at x = a and left-continuous at x=b.

Example: Show that there is some u with 0 < u< 2 such that u2 + cos(pu) = 4.

To do this we apply the IVT to the function $h(u) = u^2 +$ cos(p u). h(u) is the sum of two functions which we know are continuous (these are Field Guide functions), so h(u) is continuous.

We can compute that:

$h(0) = \quad 0^2 + \cos(0p) = 1.$

$h(2) = \quad 2^2 + \cos(2p) = 5.$

Since h(0) = 1 < 4 < 5 = h(2), and since h is continuous, the IVT implies that h(u) = 4 for some u in the closed interval [0,2]. This u cannot be one of the endpoints, so 0 < u< 2.

INTERMEDIATE ZERO THEOREM

A useful special case of the Intermediate Value Theorem is called the Intermediate Zero Theorem.

DESCRIPTION

Let f(x) be a function which is continuous on the closed interval [a,b] and suppose that f(a) and f(b) have opposite signs or that one is zero. (This can be expressed by saying that f(a)f(b) $\leq$ 0.). Then there is at least one c with $a \leq c \leq b$ such that f(c) = 0.

Example: Let $f(x) = x^2 - 2$. Then f(1) = –1 and f(2) = 2. It follows that there is some c between 1 and 2 such that f(c) = 0. If f(c) = 0, then $c^2 = 2$. Thus we have shown that 1 < sqrt(2) < 2. In Stage 2 we argued that 9 = sqrt(81) < sqrt(99) < sqrt(100) = 10.

We can now reach the same conclusion about the placement of sqrt(99) by applying the Intermediate Zero Theorem to the function s(x) = x2 - 99. Since s(9) < 0 and s(10)

> 0, we conclude that s has a root in between 9 and 10. Knowing that there is a unique positive square root for a positive real number, we conclude that 9 < sqrt(99) < 10.

APPLICATIONS OF INTERMEDIATE ZERO THEOREM

Finding Roots

Remember the calculator Aunt Emmy gave you? On that calculator is a "sqrt" button. When you punched in 99 the display read,

9 . 9 4 9 8 7

How did the calculator do this calculation? That is a pretty good question, and the answer is that very sophisticated methods, based on Calculus, have been developed for efficiently doing these types of computations. One of the underlying ideas for finding square roots and other roots of equations is called the bisection method. The bisection method is an application of the Intermediate Zero Theorem.

ODD-DEGREE POLYNOMIALS

Another type of application of the Intermediate Zero Theorem is not to find a root but to simply show that a root exists.

Example: A statement which might be familiar is the statement that every odd degree polynomial p(x) has at least one real root. This is another consequence of the Intermediate Zero Theorem, the fact that polynomial functions are continuous, and the fact that odd degree polynomials are sometimes positive and sometimes negative. Once we can establish that p(x) < 0 for some x and that p(x) > zero for some other x, then we can immediately conclude that p(x) has a root somewhere in between.

EXTREME VALUE THEOREM

DEFINITION

The Extreme Value Theorem, sometimes abbreviated EVT, says that a continuous function has a largest and smallest value on a closed interval.

This is used to show thing like:

- There is a way to set the price of an item so as to maximize profits.
- Among all ellipses enclosing a fixed area there is one with a smallest perimeter. (The circle, in fact.)
- What goes up must come down.

Like the IVT, the EVT tells us that something exists but does not say where. It is of some comfort to know that we can maximize profits, but the shrewd merchant wants to know how to set the price. Calculus gives us methods to solve problems of this nature and these methods are the subject of Stage 8.

For now, we should understand what the Extreme Value Theorem says and the role of BOTH hypotheses: that the function be continuous and that the interval be closed.

EXAMPLES

The following three examples illustrate:

1. A standard application of the Extreme Value Theorem.
2. How the Extreme Value Theorem can fail if the interval is not closed, and
3. How the Extreme Value Theorem can fail if the function is not continuous.

Example 1: The function sin(x) - cos(x) has a largest value on the interval [0, 2p].

Justification: sin(x) - cos(x) is a continuous function and [0,2p] is a closed interval. Where does this maximum occur and what is the maximum value? These questions are not answered by the EVT. We can get approximate answers by examining the graph.

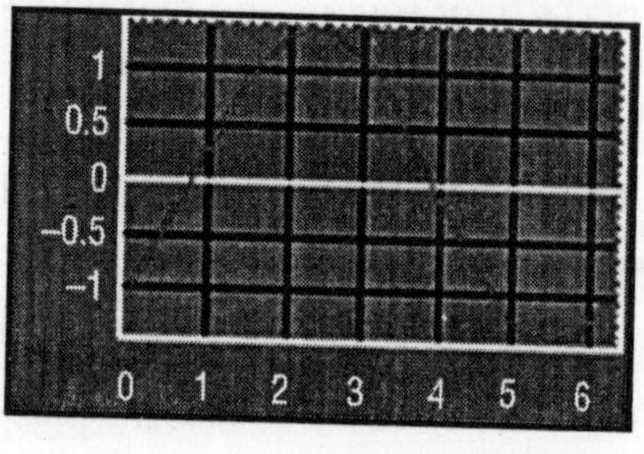

Example 2: The function f(x) = x does NOT attain a minimum value on the interval (0,1].

Explanation: The reason that f, which is a continuous function, can get away with not attaining a minimum value is that the interval (0,1] is not closed. If zero were in the interval, then f(0)=0 would be the minimum value. To formally show that there can be no minimum value we argue by contradiction.

Suppose ewere the minimum value. Since f(x) is the identity function, ewould then be an element of the interval (0,1], so e> 0. But f(e/2) = e/2 is also a value of f(x) on (0,1]. Since e/2 < e, this contradicts the assumption that ewas the minimum value of f(x) on (0,1].

Example 3. Let h(x) be the piecewise function

$$h(x)\begin{cases} 1/x x\neq 0 \\ 0 0 \end{cases} x =$$

Then h(x) is defined everywhere but is not continuous at x=0. Now consider h(x) on the interval [–1,1]. We have that

lim h(x) = ∞ and lim h(x) = $-\infty$

x --> 0+ x --> 0-

It follows that h(x) assumes both arbitrarily large and arbitrarily negative values on the interval [–1,1], so it has neither a maximum nor a minumum on this closed interval. This shows the role which continuity plays in the EVT.

RATIONAL FUNCTIONS

Rational functions are continuous for all real numbers except at those where the denominator is zero. If the denominator of a rational function f(x) is zero at x=a, then it contains some number of factors of (x – a).

- If f(x) has more factors of (x – a) in the denominator than the in the numerator, then f(x) has an asymptote at x=a. In this case the f(x) has an essential discontinuity at x=a.
- If f(x) has at least as many factors of (x–a) in the numerator as in the denominator, then f(x) is not defined at x=a but has a removable discontinuity at x=a. The value f(a) can be defined by cancelling all of the factors of (x – a) in the denominator and then

evaluating at x=a. This is the definition of f(a) which makes f(x) contininous at x=a.

Example: The numerator and denominator of the rational function $x^7 + 6x^6 + 6x^5 - 12x^4 - 19x^3 - 90x^2 - 108x + 216$ h(x) = $x^5 + x^4 - 13x^3 - x^2 + 48x - 36$ factor as,

$$(x - 2)(x + 3)^3 (x^2 + 4)(x - 1)$$
$$h(x) = \quad (x - 2)^2 (x + 3)^2 (x - 1)$$

Therefore h(x) is continuous at all points except x=1, x = 2, and x = –3. We can re-write the function h(x) as,

$$(x - 2)(x + 3)^3 (x^2 + 4)(x - 1)(x + 3)(x^2 + 4)$$
$$h(x) = \quad x \neq 1, 2, -3.$$
$$(x - 2)^2 (x + 3)^2 (x - 1)(x - 2)$$

The discontinuity at x =1 is removable and setting,

(1 + 3) (12 + 4), h(1) = –20, (1 - 2)

makes h(x) continuous at x = 1. At x = 2 the function h (x) has an asymptote and hence an essential singularity. At x = –3 the function h (x) has a removable singularity and setting h(3) = 0 makes h continuous at x = –3.

Equipment Check: There are three points at which the function,

$$x^3 + 3x^2 - 4$$
$$g(x) = x^3 - x^2 - 4x + 4$$

is not defined. Enter one of these points in the first box and then choose whether the corresponding discontinuity is essential or removable. Check your answer by pressing the "check answer" button. Repeat this for each of the three points.

There are two "help" buttons: The first will display the factored form of g(x) and the second explains all of the answers at once. Use the first only if you are stuck and the second only after trying to get the answers yourself.

7

Differential Curves and Surfaces

RIEMANN SURFACES

Riemann Surfaces were introduced in the 19th century to help resolve strange formulae and explain mysterious behaviour arising from problems in function theory. Riemann's initial discription was confusing to many people and the attempt to understand these objects and his wonderful new ideas led to the growth or creation of many fields of mathematics: topology, the theory of manifolds, potential theory, and algebra to name a few.

The idea of a Riemann surface is now a central one in mathematics, and appears in such seemingly diverse areas as integrable systems, number theory, algebraic geometry, and string theory. Because of the many points of view on Riemann surfaces, and the large number of other parts of mathematics that they interact with, there are a lot of different approaches, and topics we could cover.The goal is to touch on several of these areas and show how they relate. In particular, it would concentrate on the interaction between the complex analytic, classical geometric, and commutative algebraic points of view.

The organization of the course will largely follow the natural flow of ideas between these subjects. We will first try to understand Riemann surfaces as complex analytic manifolds. We want to understand the basic topological properties of these surfaces, and of the maps between them. We will then look at curves as submanifolds of complex manifolds, in particular as submanifolds of the projective plane. This will lead to the beautiful classical theorems in the plane: Bezout's theorem, and the adjunction formula. The description as subvarieties of the projective plane will allow us to treat the curves purely algebraically. We'll study how to describe the curves using only algebraic objects like rings and ideals, and how to express maps between curves in the language of rings. Finally, we'll largely return to the manifold point of view and understand the final fundamental theorem about algebraic curves: the Riemann-Roch theorem, which governs the geometry of curves in subtle ways. We will also look at applications of this theorem to geometric questions.

CURVES

Curves that can be given in implicit form as $f(x,y) = 0$, where f is a polynomial, are called algebraic. The degree of f is called the degree or order of the curve. Thus conics are algebraic curves of degree two. Curves of degree three already have a great variety of shapes, and only a few common ones will be given here. The simplest case is when the curve is the graph of a polynomial of degree three: $y = ax^2 + bx^2 + cx + d$, with $a \neq 0$. This curve is a (general) cubic parabola. It is symmetric with respect to the point B where $x = -b/3a$.

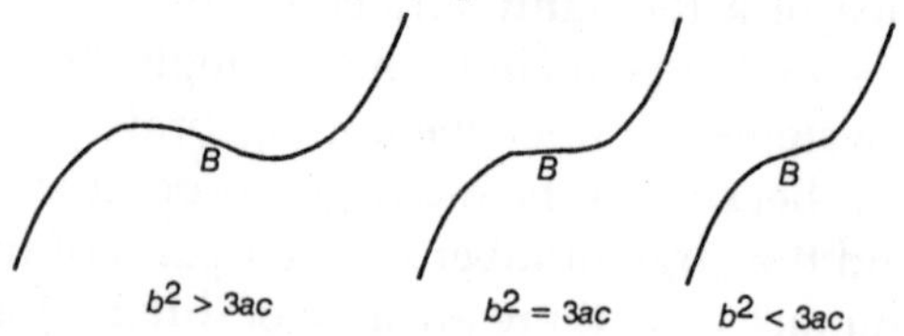

Fig. The General Cubic Parabola for $a > 0$.
For $a < 0$, Reflect in a Horizontal Line

The semicubic parabola has equation $y^2 = kx^3$; by proportional scaling one can take $k = 1$. The curve should not be confused with the cissoid of Diocles, which has equation $(a - x)y^2 = x^3$ with $a \neq 0$. The latter is asymptotic to the line $x = a$, while the semicubic parabola has no asymptotes. The cissoid's points are characterized by the equality $OP = AB$. One can take $a = 1$ by proportional scaling. More generally, any curve of degree three with equation $(x - x^0)y^2 = f(x)$, where f is a polynomial, is symmetric with respect to the x-axis and asymptotic to the line $x = x^2$.

In addition to the cissoid, the following particular cases are important:

1. The witch of Agnesi has equation $xy^2 = a^2 (a - x)$, with $a \neq 0$, and is characterized by the geometric property, right. The same property provides the parametric representation $x = a(1 + \sin 2 q)$, $y = a \tan q$. Once more, proportional scaling reduces to the case $a = 1$.
2. The folium of Descartes has equation $(x-a)y^2 = -x^2 (1/3\ x + a)$, with $a \neq 0$ (reducible to $a = 1$ by proportional scaling). By rotating 135°(right) we get the alternative and more familiar equation $x^3 + y^3 = cxy$, where $c = 1/3\sqrt{2}\ a$. The folium of Descartes is a rational curve, this is, it has a parametric representation by rational functions. In the tilted position such a representation is $x = ct/(1 + t^3)$, $y = ct^2/(1 + t^3)$ (so that $t = y/x$).
3. The strophoid has equation $(x - a)y^2 = - x^2 (x + a)$, with $a \neq 0$ (reducible to $a = 1$ by proportional scaling). It satisfies the property $AP=AP'= OA$, right; this means that POP' is a right angle. The strophoid has the polar representation $r = - a \cos 2\theta \sec \theta$, and the rational parametric representation $x = a(t^2 - 1)/(t^2 +1)$, $y = at (t^2 - 1)/(t^2 + 1)$ (so that $t = y/x$).

Among the important curves of degree four are the following:

1. A Cassini's oval is characterized by the following condition: given two foci F and $F¢$, a distance $2a$ apart, a point P belongs to the curve if the product of the distances PF and $PF¢$ is a constant k^2. If the foci are on the x-axis and equidistant from the origin, the curve's equation is $(x^2 + y^2 + a^2) - 4a^2 x^2 = k^4$.

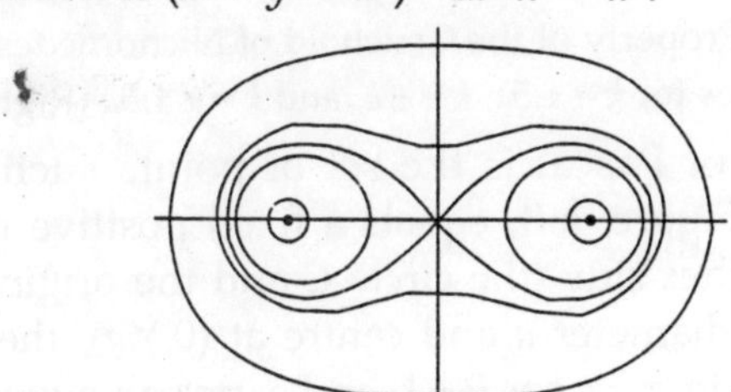

Fig. Cassini's Ovals for $k = 5a.9a$, $1.1a$ and $1.5a$ (from the Inside to the Outside).

The Foci (Dots) are *at* $x = a$ and $x = -a$. The Black Curve, $k = a$, is also Called Bernoulli's Lemniscate

Changes in a correspond to rescaling, while the value of k/a controls the shape: the curve has one smooth piece, one piece with a self-intersection, or two pieces depending on whether k is greater than, equal to, or smaller than Figure. The case $k = a$ is also known as the lemniscate (of Jakob Bernoulli); the equation reduces to $(x^2 + y^2)^2 = a^2 (x^2 - y^2)$, and upon a 45° rotation to $(x^2 + y^2)^2 = 2a\,xy$. Each Cassini's oval is the section of a torus of revolution by a plane parallel to the axis of revolution.

2. A conchoid of Nichomedes is the set of points such that the signed distance AP in Figure, left, equals a fixed real number k (the line L and the origin O being fixed). If L is the line $x = a$, the conchoid has polar equation

$$r = a \sec \theta + k.$$

Once more, a is a scaling parameter, and the value of k/a controls the shape: when $k > -a$ the curve is smooth, when $k = -a$ there is a cusp, and when $k < -a$ there is a self-intersection. The curves for k and $-k$ can also be considered two leaves of the same conchoid, with cartesian equation $(x - a)^2 (x^2 + y^2) = k^2 x^2$.

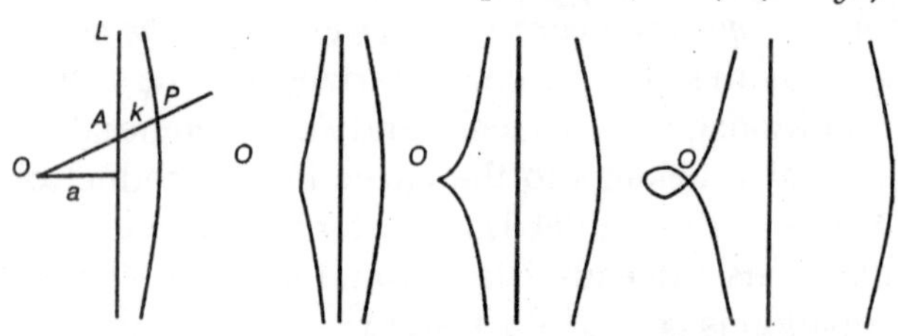

Fig: Defining Property of the Conchoid of Nichomedes (Left), and Curves for $k = \pm.5a$, $k = \pm a$, and $k = \pm 1.5a$ (Right)

A limaçon of Pascal is the set of points such that the distance AP in Figure, left, equals a fixed positive number k measured on either side (the circle C and the origin O being fixed). If C has diameter a and centre at $(0, \frac{1}{2}a)$, the limaçon has polar equation, $r = a \cos \theta + k$, and cartesian equation $(x^2 + y^2 - ax)^2 = k^2 (x^2 + y^2)$. The value of k/a controls the shape, and there are two particularly interesting cases.

For $k = a$ we get a cardioid. For $a = \frac{1}{2}k$ we get a curve that can be used to trisect an arbitrary angle α : if we draw a line L through the centre of the circle C and making an angle α with the positive x-axis, and if we call P the intersection of L with the limaçon $a = 2k$, the line from O to P makes an angle $1/3\,\alpha$ with L.

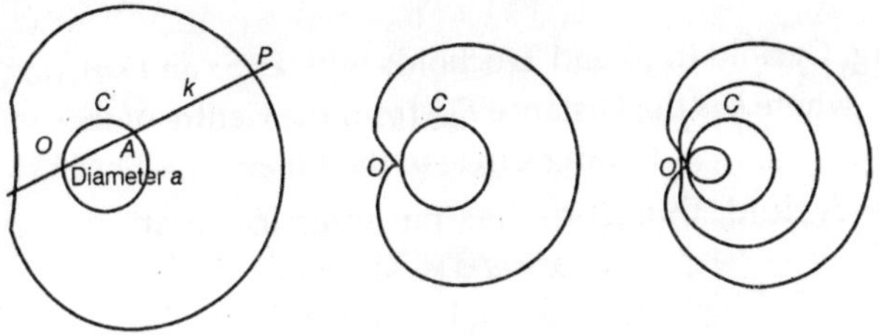

Fig. Defining Property of the Limaçon of Pascal (left), and curves for k=1.5a, k=a, and k=.5a (right). The Middle curve is the Cardioid, the one on the right a Trisectrix

Hypocycloids and epicycloids with rational ratio are also algebraic curves, generally of higher degree.

ROULETTES (SPIROGRAPH CURVES)

Suppose given a fixed curve C and a moving curve M, which rolls on C without slipping. The curve drawn by a point P kept fixed with respect to M is called a roulette, of which P is the pole.

The most important examples of roulettes arise when M is a circle and C is a straight line or a circle, but an interesting additional example is provided by the catenary $y = a \cosh(x/a)$, which arises by rolling the parabola $y = x^2/(4a)$ on the x-axis with pole the focus of the parabola (that is, $P = (0,a)$ in the initial position).

The catenary is the shape taken under the action of gravity by a chain or string of uniform density whose ends are held in the air. A circle rolling on a straight line gives a trochoid, with the cycloid as a special case when the pole P lies on the circle. If the moving circle M has radius a and the distance from the pole P to the centre of M is k, the trochoid has the parametric equation,

$$x = a\varphi - k \sin\varphi,\ y = a - k \cos\varphi.$$

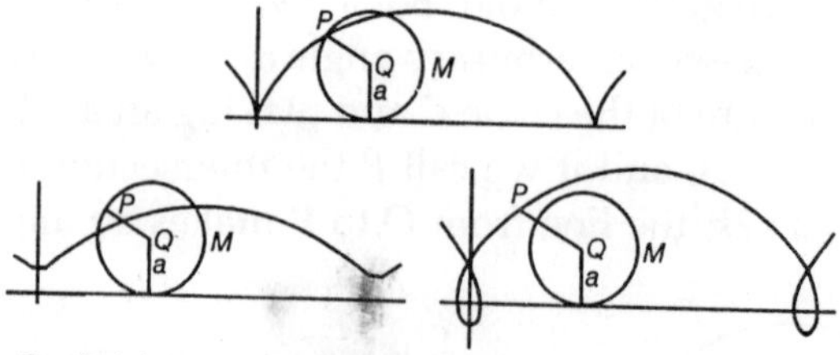

Fig. Cycloid (top) and Trochoids with k=.5a and k=1.5a, where k is the Distance PQ from the Centre of the Rolling Circle to the Pole

The cycloid, therefore, has parametric equation,

$$x = a\varphi(-\sin\varphi),$$
$$y = a(1-\cos\varphi).$$

One can eliminate φ to get x as a (multivalued) function of y, which takes the following form for the cycloid:

$$x = \text{arc cos}\ \frac{a-y}{a} \pm \hat{G}$$

The length of one arch of the cycloid is $8a$, and the area under the arch is $3\pi a^2$. A trochoid is also called a curtate cycloid when $k<a$ (that is, when P is inside the circle) and a prolate cycloid when $k>a$. A circle rolling on another circle and exterior to it gives an epitrochoid.

If a is the radius of the fixed circle, b that of the rolling circle and k is the distance from P to the centre of the rolling circle, the parametric equation of the epitrochoid is,

$$x = (a+b)\cos\theta - k\cos((1+a/b)\theta),$$
$$y = (a+b)\sin\theta - k\sin((1+a/b)\theta).$$

These equations assume that at the start everything is aligned along the positive x-axis, left.

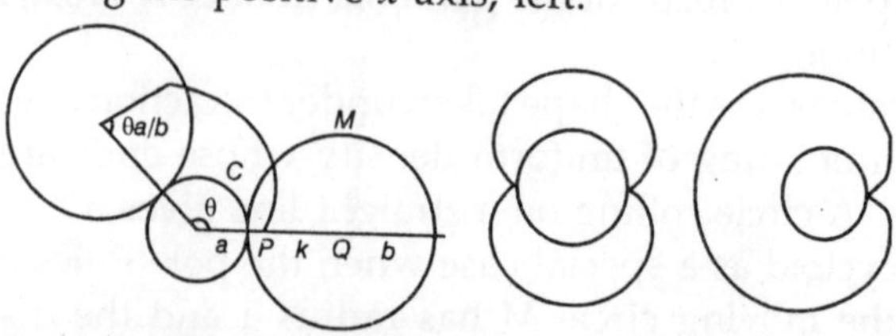

- *Left:* Initial Configuration for Epicycloid (black) and Configuration at Parameter value (red).
- *Middle*: Epicycloid with $b = ½a$ (nephroid).
- *Right*: Epicycloid with $b = a$ (Cardioid).

Usually one considers the case when a/b is a rational number, say $a/b = p/q$ where p and q are relatively prime. Then the rolling circle returns to its original position after rotating q times around the fixed circle, and the epitrochoid is a closed curve---in fact, an algebraic curve. One also usually takes $k = b$, so that P lies on the rolling circle; the curve in this case is called an epicycloid. The middle diagram shows the case $b = k = \frac{1}{2}a$, called the nephroid; this curve is the cross section of the caustic of a spherical mirror. The diagram on the right shows the case $b = k = a$, which gives the cardioid. Hypotrochoids and hypocycloids are defined in the same way as epitrochoids and epicycloids, but the rolling circle is inside the fixed one.

The parametric equation of the hypotrochoid is,

$$x = (a - b)\cos\theta + k\cos((a/b - 1)\theta),$$

$$y = (a - b)\sin\theta - k\sin((a/b - 1)\theta),$$

where the letters have the same meaning as for the epitrochoid. Usually one takes a/b rational and $k = b$.

There are several interesting particular cases:

1. $b = k = a$ gives a point.
2. $b = k = \frac{1}{2}a$ gives a diameter of the circle C.
3. $b = k = a$ gives the deltoid which has the algebraic equation, $(x^2 + y^2)^2 - 8ax + 24axy^2 + 18a(x^2 + y^2) - 27a^4 = 0$.
4. $b = k = \frac{1}{4}a$ gives the astroid, an algebraic curve of degree six whose equation can be reduced to $x^{2/3} + y^{2/3} = a^{2/3}$. The figure illustrates another property of the astroid: its tangent intersects the coordinate axes at points that are always the same distance a apart. Otherwise said, the astroid is the envelope of a moving segment of fixed length whose endpoints are constrained to lie on the two coordinate axes.

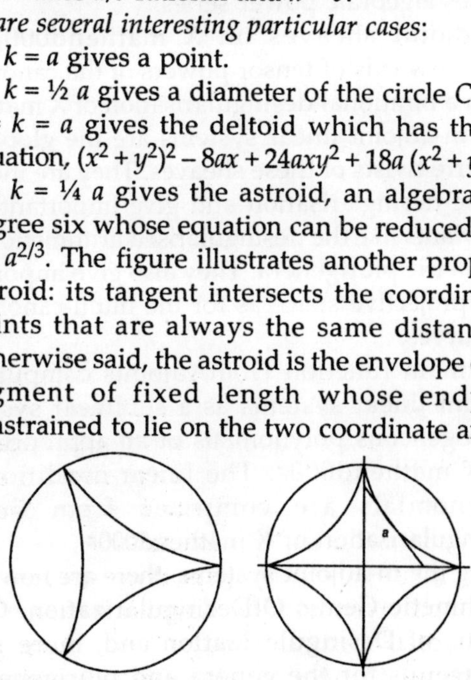

Fig. The Hypocycloids with $a = 3b$ (Deltoid) and $a = 4b$ (Astroid)

ALGEBRAIC SURFACES FEATURES

A new package by Töbias Beck and Joseph Schicho for working with algebraic hypersurfaces in 3-dimensional projective space has been added. This has three main components: formal desingularisation, computation of m-adjoint linear systems and full classification and parametrisation of rational hypersurfaces over Q mathend.

- Formal Desingularisation of X mathend000# (main function ResolveProjectiveSurface) computes the power series completions of the generic rings of the components of some (not necessarily minimal) birational desingularisation of X mathend000# that lie over components of the singular subscheme of X mathend000#. This is based on a Jung-Hirzebruch desingularisation of X mathend000# and the algorithm was designed and implemented by T. Beck in his PhD. It uses algebraic power series.
- m-adjoint sheaves on X mathend000# are the pushforwards of tensor powers of the canonical sheaf from a birational desingularisation of X mathend000#. The m-adjoint linear systems are the global sections of Serre twists of these sheaves. They are independent of the desinglarisation and give important birational invariants like the desingularised arithmetic genus and geometric plurigenera. They also give important maps into projective space as for the multi-canonical maps on curves.
- The main function HomAdjoints computes the m-adjoint linear systems as a sublinear system of all homogeneous polynomials of an appropriate degree on X mathend000#. The linear conditions on the polynomials are computed from the formal desingularisation of X mathend000#.
- Using the m-adjoint systems, there are now functions Arithmetic Genus OfDesingularization, Geometric Genus of Desingularization and, more generally, Plurigenus for the genera and plurigenera of any desingularisation of X mathend000#. There is also a

function IsRational to determine whether X mathend000# is rational using Castelnuevo's criterion on a desinglarisation. This says that X mathend000# is rational iff the arithmetic genus and second plurigenus of a desingularisation are both zero.

- For a rational hypersurface X mathend000#, a well-chosen m-adjoint map will take X mathend000# to a special rational surface nicely embedded in projective space (usually a ruled or Del Pezzo surface). There is a function Classify that finds such an adjoint map, following an algorithm of Josef Schicho, and the special surface image of X mathend000# in one of the families of Schicho's classification.
- Finally, there is a suite of functions to parametrise these special surfaces. These are all based on algorithms of Schicho and were implemented by him and Beck. The existing Magma functions for non-singular degree 6,8 and 9 Del Pezzo surfaces are used. The rest are new. They cover quadrics in P3 mathend000#, ruled surfaces, Del Pezzo surfaces of degree 7 and singular Del Pezzos of degrees 6 and 8 and functions to reduce low degree Del Pezzos to degree 5 or higher. The only case currently missing is for degree 5 Del Pezzos and this will be filled with code from Josef Schicho in the near future.

ABOUT SPIRALS

A number of interesting curves have polar equation $r = f(\theta)$, where f is a monotonic function (always increasing or decreasing).

This property leads to a spiral shape. The logarithmic spiral or Bernoulli spiral is self-similar: by rotation the curve can be made to match any scaled copy of itself. Its equation is $r = k^{2\theta}$; the angle between the radius from the origin and the tangent to the curve is constant, and equal to φ = arccot a. A curve parametrized by arclength and such that the radius curvature is proportional to the parameter at each point is a Bernoulli spiral.

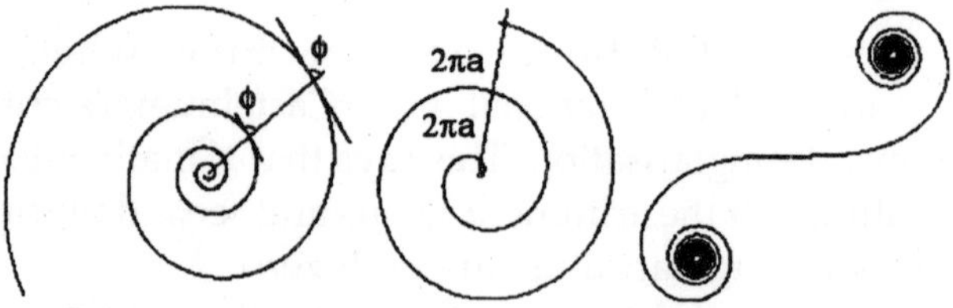

Fig. The Bernoulli or Logarithmic Spiral (left), the Archimedes or Linear S Piral (Middle) and the Cornu Spiral (Right)

In the Archimedean spiral or linear spiral, it is the spacing between intersections along a ray from the origin that is constant. The equation of this spiral is $r = a\theta$; by scaling one can take $a = 1$. It has an inner endpoint, in contrast with the logarithmic spiral, which spirals down to the origin without reaching it. The Cornu spiral or clothoid important in optics and engineering, has the following parametric representation in Cartesian coordinates:

$$x = aC\,(t) = a\int_0^t \cos\left(\frac{1}{2}\pi s^2\right)ds,\ y = aS\,(t) = a\int_0^t \sin\left(\frac{1}{2}\pi s^2\right)ds,$$

(C and S are the so-called Fresnel integrals) A curve parametrized by arclength and such that the radius curvature is inversely proportional to the parameter at each point is a Cornu spiral (compare the Bernoulli spiral).

PEANO AND FRACTAL CURVES

There are examples of curves (in the sense of continuous maps from the real line to the plane) that completely cover a two-dimensional region of the plane. We give a construction of such a Peano curve, adapted from David Hilbert's example. The construction is inductive, and is based on replacement rules. We consider building blocks of six shapes,

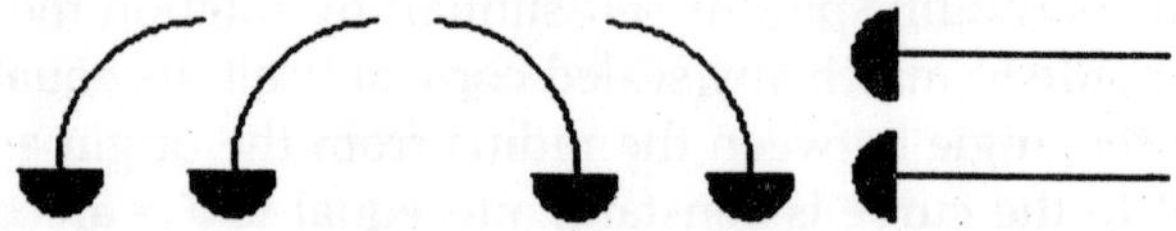

Fig. The Plano Curve

the length of the straight pieces being twice the radius of the curved ones. A sequence of these patterns end-to-end represents a curve, if we disregard the red and green half-disks.

The replacement rules are the following:

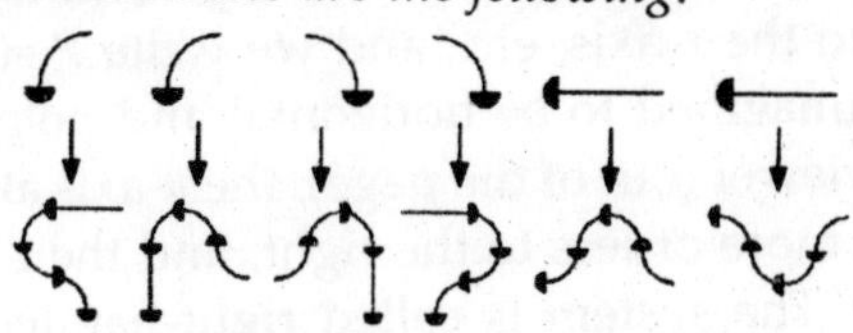

Fig. Fractal Curves

The rules are applied taking into account the way each piece is turned. Here we apply the replacement rules to a particular initial pattern:

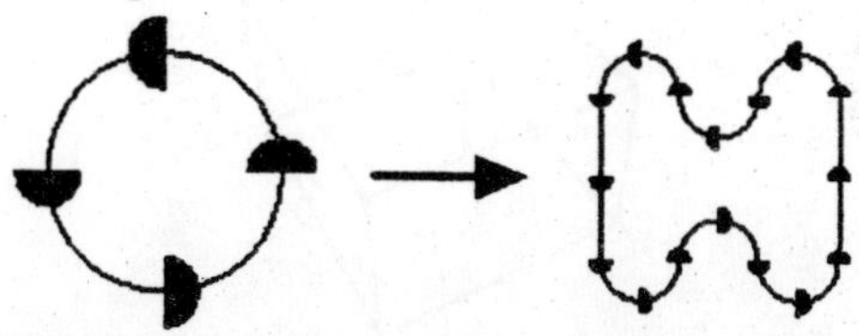

(We scale the result so it has the same size as the original.) Applying the process repeatedly gives, in the limit, the Peano curve.

Here are the first five steps:

The same idea of replacement rules leads to many interesting fractal, and often self-similar, curves.

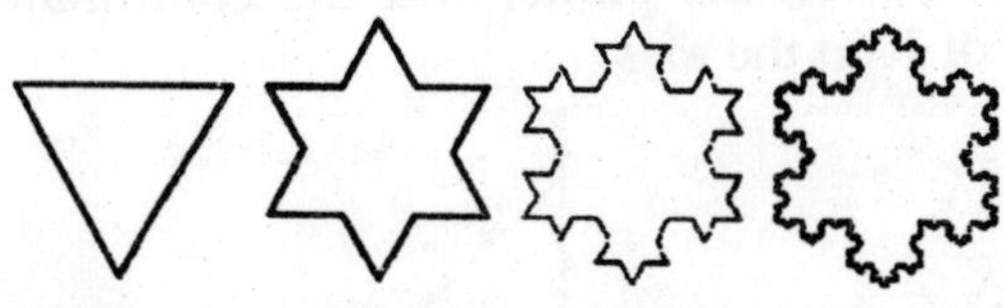

For example, the substitution leads to the Koch snowflake when applied to an initial equilateral triangle, like this (the first three stages and the sixth are shown):

CARTESIAN COORDINATES

In cartesian coordinates (or rectangular coordinates), a point P is referred to by three real numbers, indicating the positions of the perpendicular projections from the point to three fixed, perpendicular, graduated lines, called the axes.

If the coordinates are denoted x, y, z, in that order, the axes are called the x-axis, etc., and we write $P = (x,y,z)$. Often the x-axis is imagined to be horizontal and pointing roughly towards the viewer (out of the page), the y-axis also horizontal and pointing more or less to the right, and the z-axis vertical, pointing up. The system is called right-handed if it can be rotated so the three axes are in this position. The point $x = 0$, $y = 0$, $z = 0$ is the origin, where the three axes intersect.

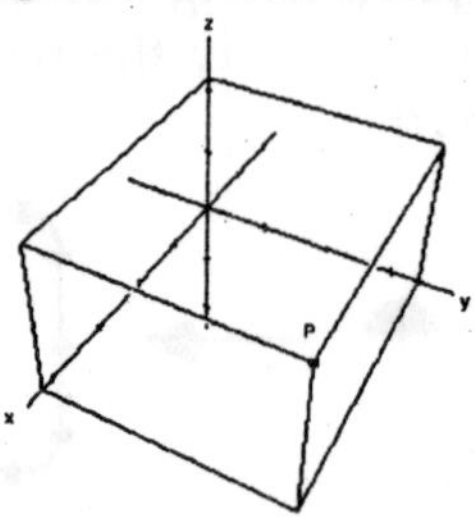

Fig. In Cartesian Coordinates, $P = (4.2,3.4,2.2)$

CYLINDRICAL COORDINATES

To define cylindrical coordinates, we take an axis (usually called the z-axis) and a perpendicular plane, on which we choose a ray (the initial ray) originating at the intersection of the plane and the axis (the origin). The coordinates of a point P are: the polar coordinates (r,θ) of the projection of P on the plane, and the coordinate z of the projection of P on the axis.

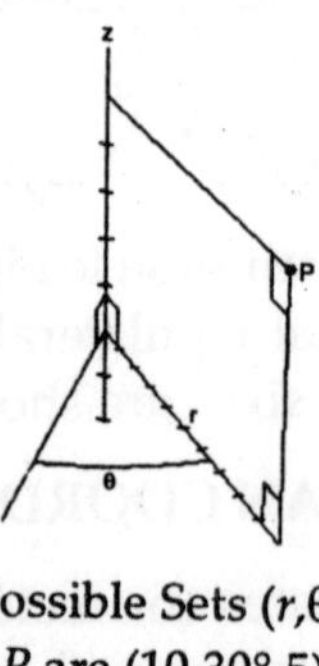

Fig. Among the Possible Sets (r,θ,z) of Cylindrical Coordinates for P are $(10,30°,5)$ and $(10,390°,5)$

SPHERICAL COORDINATES

To define spherical coordinates, we take an axis (the polar axis) and a perpendicular plane (the equatorial plane), on which we choose a ray (the initial ray) originating at the intersection of the plane and the axis (the origin *O*).

The coordinates of a point *P* are: the distance θ from *P* to the origin; the angle φ (zenith) between the line *OP* and the positive polar axis; and the angle θ (azimuth) between the initial ray and the projection of *OP* to the equatorial plane. As in the case of polar and cylindrical coordinates, θ is only defined up to multiples of 360°, and likewise φ.

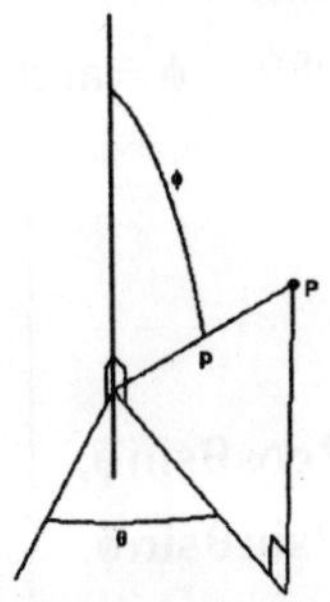

Fig. A set of Spherical Coordinates for *P* is (*P*,θ, φ) = (10,60°,30°)

Usually φ is assigned a value between 0 and 180°, but values of φ between 180° and 360° can also be used; the triples (*P*, φ ,θ) and (*P*, 360°– φ , 180°+θ) represent the same point. Similarly, one can extend *P* to negative values; the triples (*P*, φ ,θ) and (-*P*, 180° – φ , 180° + θ) represent the same point.

RELATION AMONG COORDINATES

Consider a cartesian, a cylindrical, and a spherical coordinate system, related as shown in Figure.

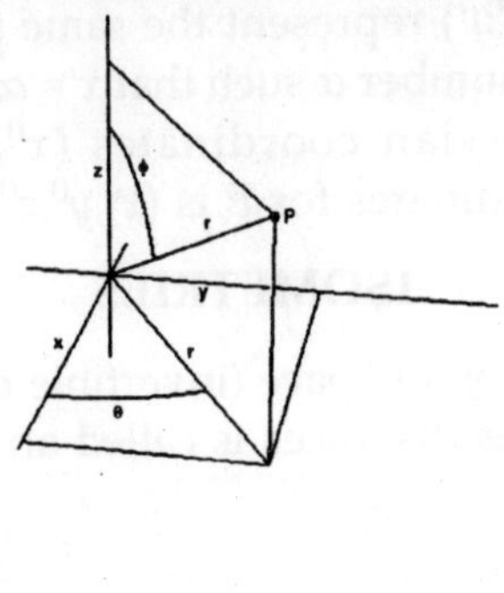

Then the cartesian coordinates (x,y,z), the cylindrical coordinates (r,θ,z), and the spherical coordinates (P,φ,θ) of a point are related as follows:

$$\text{cart} \leftrightarrow \text{cyl} \begin{cases} x = r\cos\theta, \\ y = r\sin\theta, \end{cases} \begin{cases} r = \sqrt{x^2+y^2}, \\ \theta = \arctan\dfrac{y}{x}, \end{cases} \begin{cases} \sin\phi = \dfrac{x}{\sqrt{r^2+z^2}}, \\ \cos\phi = \dfrac{y}{\sqrt{r^2+z^2}} \end{cases}$$

$$\text{cyl} \leftrightarrow \text{sph} \begin{cases} r = P\sin\phi, \\ z = P\cos\phi, \end{cases} \begin{cases} P = \sqrt{r^2+z^2}, \\ \phi = \arctan\dfrac{r}{z}, \end{cases} \begin{cases} \sin\phi = \dfrac{r}{\sqrt{r^2+z^2}}, \\ \cos\phi = \dfrac{z}{\sqrt{r^2+z^2}} \end{cases}$$

$$\text{cart} \leftrightarrow \text{sph} \begin{cases} x = P\cos\theta\sin\phi, \\ y = P\sin\theta\sin\phi, \\ z = P\cos\phi, \end{cases} \begin{cases} P = \sqrt{x^2+y^2+z^2} \\ \theta = \arctan\dfrac{y}{x}, \\ \phi = \arctan\dfrac{\sqrt{x^2+y^2}}{z} \\ = \arccos\dfrac{z}{\sqrt{x^2+y^2+z^2}}. \end{cases}$$

HOMOGENEOUS COORDINATES

A quadruple of real numbers $(x{:}y{:}z{:}t)$, with $t \neq 0$, is a set of homogeneous coordinates for the point P with cartesian coordinates $(x/t, y/t, z/t)$.

Thus the same point has many sets of homogeneous coordinates: $(x{:}y{:}z{:}t)$ and $(x'{:}y'{:}z'{:}t')$ represent the same point if and only if there is some real number α such that $x' = \alpha x$, $y' = \alpha y$, $z' = \alpha z$, $t' = \alpha t$. If P has cartesian coordinates (x^0,y^0,z^0), one set of homogeneous coordinates for P is $(x^0,y^0,z^0,1)$.

ISOMETRIES

A transformation of space (invertible map of the plane to itself) that preserves distances is called an isometry of space.

Every isometry of space is of one of the following types:

- The identity (which leaves every point fixed);
- A translation by a vector v;
- A rotation through an angle a around a line L;
- A screw motion through an angle α around a line L, with displacement d;
- A reflection in a line P;
- A glide-reflection in a line P with displacement vector v.

The identity is a particular case of a translation and of a rotation; rotations are particular cases of screw motions; and reflections are particular cases of glide-reflections. However, as in the plane case, is more intuitive to consider each case separately.

FORMULAS FOR SYMMETRIES

In the formulas below, a multiplication between a matrix and a triple of coordinates should be carried out regarding the triple as a column vector (or a matrix with three rows and one column).

Translation by, (x^0,y^0,z^0): $(x,y,z) \mapsto (x+x^0,\ y+y^0,\ z+z^0)$

Rotation through α (counterclockwise) around the line through the origin with direction cosines a, b, c:

$$(x,y,z) \mapsto M(x,y,z),$$

where M is the matrix,

$$\begin{bmatrix} a^2(1-\cos\alpha)+\cos\alpha & ab(1-\cos\alpha)-c\sin\alpha & ac(1-\cos\alpha)+b\sin\alpha, \\ ab(1-\cos\alpha)+c\sin\alpha & b^2(1-\cos\alpha)+\cos\alpha & bc(1-\cos\alpha)-a\sin\alpha \\ ac(1-\cos\alpha)-b\sin\alpha & bc(1-\cos\alpha)+a\sin\alpha & c^2(1-\cos\alpha)+\cos\alpha \end{bmatrix}$$

Rotation through α (counterclockwise) around the line with direction cosines a,b,c through an arbitrary point (x^0,y^0,z^0):

$$(x,y,z) \_ (x^0,y^0,z^0)+M(x-x^0,\ y-y^0,\ z-z^0),$$

where M is given by (1).

Arbitrary rotations and Euler angles. Any rotation of space fixing the origin can be decomposed as a rotation by φ about the z-axis, followed by a rotation by θ about the y-axis, followed by rotation by about the z-axis. The numbers φ, θ and ψ are called the Euler angles of the composite rotation, which acts as follows:

$$(x,y,z) \mapsto M(x,y,z),$$

where M is the matrix given by,

$$\begin{bmatrix} \cos\phi\cos\theta\cos\psi - \sin\phi\sin\psi & -\sin\phi\cos\theta\cos\psi - \cos\phi\sin\psi & \sin\theta\cos\psi \\ \sin\phi\cos\psi + \cos\phi\cos\theta\sin\psi & -\sin\phi\cos\theta\sin\psi + \cos\phi\cos\psi & \sin\theta\sin\psi \\ -\cos\phi\sin\theta & \sin\theta\sin\phi & \cos\theta \end{bmatrix}$$

(An alternative decomposition, more natural if we think of the coordinate system as a rigid trihedron that rotates in space, is the following: a rotation by ψ about the z-axis, followed by a rotation by θ about the rotated y-axis, followed by a rotation by φ about the rotated z-axis. Note that the order is reversed.) Provided that θ is not a multiple of 180°, the decomposition of a rotation in this form is unique (apart from the ambiguity arising from the possibility of adding a multiple of 360°to any angle). Figure shows how the Euler angles can be read off geometrically.

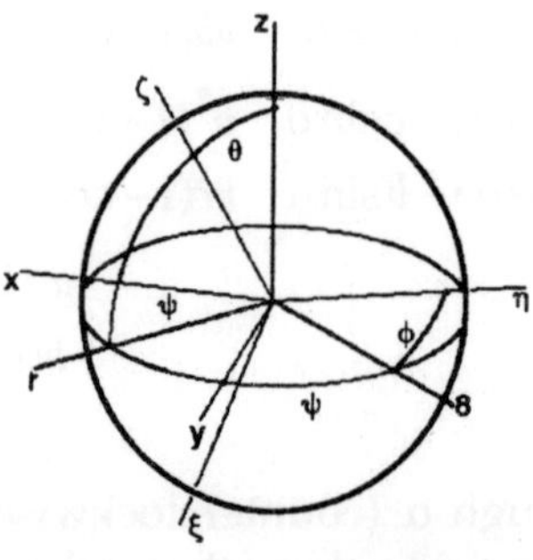

Fig. The coordinate rays Ox, Oy, Oz, Together with their Images $O\xi$, $O\eta$, $O\zeta$ under a rotation, fix the Euler angles associated with that Rotation, as follows: $\theta = zO\zeta$, $\psi = xOr = y\,Os$, and $= sO\eta$.

Here the ray *Or* is the projection of *O* to the *xy*-plane. The ray *Os* is Determined by the Intersection of the *xy*- and -planes.

Warning: Some references define Euler angles differently; the most common variation is that the second rotation is taken about the *x*-axis instead of about the *y*-axis.

Screw motion: With angle α and displacement *d* around the line with direction cosines *a, b, c* through an arbitrary point (x^0,y^0,z^0): $(x,y,z) \mapsto (x^0 + ad,\ y^0 + bd,\ z^0 + cd) + M(x-x^0,\ y - y^0,\ z-z^0)$, where *M* is given by (1).

REFLECTION

Reflection in a plane with equation ax + by + cz + d = 0:

$$(x, y) \mapsto \frac{1}{a^2 + b^2 + c^2}\,(M(x^0, y^0, z^0) - (2ad, 2bd, 2cd)),$$

where *R* is the matrix,

$$M = \begin{pmatrix} -a^2 + b^2 + c^2 & -2ab & -2ac \\ -2ab & a^2 - b^2 + c^2 & -2bc \\ -2ac & -2bc & a^2 + b^2 - c^2 \end{pmatrix}.$$

Reflection: In a plane going through (x^0,y^0,z^0) and whose normal has direction cosines *a, b, c*:

$$(x,y) - (x^0 + y^0 + z^0) + M(x - x^0,\ y - y^0,\ z - z^0),$$

where *M* is as in (3). Glide-reflection in a plane *P* with displacement vector *v*: Apply first a reflection in *P*, then a translation by the vector *v*.

HOMOGENEOUS COORDINATES

All isometries of space can be expressed in homogeneous coordinates in terms of multiplication by a matrix. As in the case of plane isometries, this means that the successive application of transformations reduces to matrix multiplication. In the formulas below,

$$\begin{bmatrix} M & 0 \\ 0 & 1 \end{bmatrix}$$

is the 4 × 4 projective matrix obtained from the 3 × 3 matrix *M* by adding a row and a column as stated.)

Translation: by (x^0,y^0,z^0):

$$\begin{bmatrix} 1 & 0 & 0 & x_0 \\ 0 & 1 & 0 & y_0 \\ 0 & 0 & 1 & z_0 \\ 0 & 0 & 0 & 1 \end{bmatrix}$$

Rotation through the origin:

$$\begin{bmatrix} M & 0 \\ 0 & 1 \end{bmatrix}$$

where M is given in (10.1.1) or (10.1.2), as the case may be.

Reflection in a plane through the origin:

$$\begin{bmatrix} M & 0 \\ 0 & 1 \end{bmatrix},$$

where M is given in (10.1.3).

From this one can deduce all other transformations, as in the case of plane transformations.

TRANSFORMATIONS OF SPACE

A transformation of space that preserves shapes is called a similarity. Every similarity of the plane is obtained by composing a proportional scaling transformation (also known as a homothety) with an isometry. A proportional scaling transformation centered at the origin has the form,

$$(x,y,z) \mapsto (ax,ay,az),$$

where $a \neq 0$ is the scaling factor (a real number). The corresponding matrix in homogeneous coordinates is,

$$H_2 = \begin{bmatrix} a & 0 & 0 & 0 \\ 0 & 1 & 0 & 0 \\ 0 & 0 & a & 0 \\ 0 & 0 & 0 & 1 \end{bmatrix}.$$

In cylindrical coordinates, the transformation is,

$$(r,\theta,z) \mapsto (ar,\varphi,\theta az).$$

In spherical coordinates, it is,

$$(r,\varphi,\theta) \mapsto (ar,\varphi,\theta).$$

AFFINE TRANSFORMATIONS

A transformation that preserves lines and parallelism (maps parallel lines to parallel lines) is an affine transformation. There are two important particular cases of such transformations: A nonproportional scaling transformation centered at the origin has the form,

$$(x,y,z) \mapsto (ax,by,cz),$$

where a, b, $c \neq 0$ are the scaling factors (real numbers). The corresponding matrix in homogeneous coordinates is,

$$H\,a,b,c = \begin{bmatrix} a & 0 & 0 & 0 \\ 0 & b & 0 & 0 \\ 0 & 0 & c & 0 \\ 0 & 0 & 0 & 1 \end{bmatrix}.$$

A shear in the x-direction and preserving horizontal planes has the form,

$$(x, y, z) \mapsto (x + rz, y, z),$$

where r is the shearing factor. The corresponding matrix in homogeneous coordinates is,

$$S_r = \begin{bmatrix} 1 & 0 & r & 0 \\ 0 & 1 & 0 & 0 \\ 0 & 0 & 1 & 0 \\ 0 & 0 & 0 & 1 \end{bmatrix}.$$

Every affine transformation is obtained by composing a scaling transformation with an isometry, or one or two shears with a homothety and an isometry.

PROJECTIVE TRANSFORMATIONS

A transformation that maps lines to lines (but does not necessarily preserve parallelism) is a projective transformation. Any spatial projective transformation can be expressed by an invertible 4×4 matrix in homogeneous coordinates; conversely, any invertible 4×4 matrix defines a projective transformation of the plane. Projective transformations (if not affine) are not defined on all of space, but only on the complement of a plane (the missing plane is "mapped to infinity").

The following particular case is often useful, especially in computer graphics, in projecting a scene from space to the plane. Suppose an observer is at the point $E = (x^0, y^0, z^0)$ of space, looking towards the origin $O=(0,0,0)$. Let P, the screen, be the plane through O and perpendicular to the ray EO. Place a rectangular coordinate system $\xi\eta$ on P with origin at O and such that the positive η -axis lies in the half-plane determined by E and the positive z-axis of space (that is, the z-axis is pointing ``up'' as seen from E).

Then consider the transformation that associates to a point $X=(x, y, z)$ the triple (ξ, η, ξ), where (ξ, η) are the coordinates of the point where the line EX intersects P (the screen coordinates of X as seen from E), and ξ is the inverse of the signed distance from X to E along the line EO (the depth of X as seen from E). This is a projective transformation, given by the matrix,

$$\begin{bmatrix} -r^2 y_0 & r^2 x_0 & 0 & 0 \\ -r x_0 z_0 & -r y_0 z_0 & rP^2 & 0 \\ 0 & 0 & 0 & rp \\ -p x_0 & -p y_0 & -p x_0 & r^2 p \end{bmatrix}$$

with $p = \sqrt{x_0^2 + y_0^2}$ and $r = \sqrt{x_0^2 + y_0^2 + z_0^2}$.

DIRECTIONS, PLANES AND LINES

DIRECTION COSINES

Given a vector (a, b, c) in three-space, the direction cosines of this vector are,

$$\cos\alpha = \frac{a}{\sqrt{a^2+b^2+c^2}},$$

$$\cos\beta = \frac{b}{\sqrt{a^2+b^2+c^2}},$$

$$\cos\chi = \frac{c}{\sqrt{a^2+b^2+c^2}},$$

Here the direction angles α, β, χ are the angles that the vector makes with the positive x-, y- and z-axes, respectively. In formulas, it is usually the direction cosines that occur, rather than the direction angles.

We have,

$$\cos^2 \alpha + \cos^2 \beta + \cos^2 \chi = 1$$

PLANES

The (cartesian) equation of a plane is linear in the coordinates x and y, that is, of the form $ax + by + cz + d = 0$. The normal direction to this plane is (a,b,c).

The intersection of this plane with the x-axis, or x-intercept, is $x = -d/a$; the y-intercept is $y = -d/b$, and the z-intercept is $z = -d/c$.

The plane is vertical (perpendicular to the xy-plane) if $c = 0$; it is perpendicular to the x-axis if $b = c = 0$; and likewise for the other coordinates.

When $a^2 + b^2 + c^2 = 1$ and $d \leq 0$ in the equation $ax + by + cz + d = 0$, the equation is said to be in normal form. In this case d is the distance of the plane to the origin, and (a, b, c) are the direction cosines of the normal. To reduce an arbitrary equation $ax + by + cz + d = 0$ to normal form, divide by,

$$\pm\sqrt{a^2 + b^2 + c^2},$$

where the sign of the radical is chosen opposite the sign of d when $d \neq 0$, the same as the sign of c when $d = 0$ and $c \neq 0$, and the same as the sign of b otherwise.

PLANES WITH PRESCRIBED PROPERTIES

Plane through (x^0, y^0, z^0) and perpendicular to the direction (a, b, c):

$$a(x - x^0) + b(y - y^0) + c(z - z^0) = 0$$

Plane through (x^0, y^0, z^0) and parallel to the directions (a^1, b^1, c^1) and (a^2, b^2, c^2):

$$\begin{vmatrix} x - x_0 & y - y_0 & z - z_0 \\ a_1 & b_1 & c_1 \\ a_2 & b_2 & c_2 \end{vmatrix} = 0$$

Plane through (x^0, y^0, z^0) and (x^1, y^1, z^1) and parallel to the direction (a, b, c):

$$\begin{vmatrix} x-x_0 & y-y_0 & z-z_0 \\ x_1-x_0 & y_1-y_0 & z_1-z_0 \\ a & b & c \end{vmatrix} = 0$$

Plane going through (x^0,y^0,z^0), (x^1,y^1,z^1) and (x^2,y^2,z^2):

$$\begin{vmatrix} x & y & 1 \\ x_0 & y_0 & 1 \\ x_2 & y_2 & 1 \end{vmatrix} = 0 \text{ or } \begin{vmatrix} x-x_2 & y-y_0 & z-z_0 \\ x_1-x_0 & y_1-y_0 & z_1-z_0 \\ x_2-x_0 & y_2-y_0 & z_2-z_0 \end{vmatrix}$$

The last three formulas remain true in oblique coordinates. The distance from the point (x^0, y^0, z^0) to the plane $ax + by + cz + d = 0$ is,

$$\left|\frac{ax_0 + by_0 + cz_0 + d}{\sqrt{a^2 + b^2 + c^2}}\right|$$

The angle between two planes $a^0x + b^0y + c^0z + d^0 = 0$ and $a^1 x + b^1 y + c^1z + d^1 = 0$ is, arcoos,

$$\frac{a_0x_1 + b_0b_1 + c_0c_1}{\sqrt{a_0^1 + b_0^2 + c_0^2}\sqrt{a_1^2 + b_1^2 + c_1^2}}$$

In particular, the two planes are parallel when $a^0: b^0: c^0 = a^1: b^1: c^1$, and perpendicular when $a^1a^1 + b^1b^1 + c^1c^1 = 0$.

CONCURRENCE AND COPLANARITY

Four planes $a^0x + b^0y + c^0z + d^0 = 0$, $a^1x + b^1 y + c^1 z + d^1 = 0$, $a^2x + b^2y + c^2z + d^2 = 0$ and $a^3x + b^3y + c^3z + d = 0$ are concurrent if and only if,

$$\begin{vmatrix} a_0 & b_0 & c_0 & d_0 \\ a_1 & b_1 & c_1 & d_1 \\ a_2 & b_2 & c_2 & d_2 \\ a_3 & b_3 & c_3 & d_3 \end{vmatrix} = 0.$$

Four points (x^0, y^0, z^0), (x^1, y^1, z^1), (x^2, y^2, z^2) and (x^{3s}, y^3, z^3) are coplanar if and only if,

$$\begin{vmatrix} x_0 & y_0 & z_0 & 1 \\ x_1 & y_1 & z_1 & 1 \\ x_2 & y_2 & z_2 & 1 \\ x_3 & y_3 & z_3 & 1 \end{vmatrix} = 0$$

(Both of these assertions remain true in oblique coordinates.)

LINES

Two planes that are not parallel or coincident intersect in a straight line, so one can express a line by a pair of linear equations,

$$ax + by + cz + d = 0 \text{ and } a'x + b'y + c'z + d' = 0$$

such that $bc' - cb'$, $ca' - ac'$, and $ab' - ba'$ are not all zero. The line thus defined is parallel to the vector,

$$(bc' - cb', ca' - ac', ab' - ba').$$

The direction cosines of the line are those of this vector. (The direction cosines of a line are only defined up to a simultaneous change in sign, since the opposite vector still gives the same line.)

The following particular cases are important: Line through (x^0, y^0, z^0) parallel to the vector (a, b, c):

$$\frac{x - x_0}{a} = \frac{y - y_0}{b} = \frac{z - z_0}{c}$$

Line through $(x^0, y^0, z^0))$ and (x^1, y^1, z^1):

$$\frac{x - x_0}{x_1 - x_0} = \frac{y - y_0}{y_1 - y_0} = \frac{z - z_0}{z_1 - z_0}$$

This line is parallel to the vector: $(x^1 - x^0, y^1 - y^0, z^1 - z^0)$.

DISTANCES

The distance between two points in space is the length of the line segment joining them.

The distance between the points (x^0, y^0, z^0) and (x^1, y^1, z^1) is is,

$$\sqrt{(x_1 - x_0)^2 + (y_1 - y_0)^2 + (z_1 - z_0)^2}$$

The point $k\%$ of the way from $P^0 = x^0, y^0, z^0)$ to $P^1 = (x^1, y^1, z^1)$ is,

$$\left(\frac{kx_1 + (100-k)x_2}{100}, \frac{ky_1 + (100-k)y_2}{100}, \frac{kz_1 + (100-k)z_2}{100}\right)$$

(The same formula works also in oblique coordinates.) This point divides the segment P^1P^0 in the ratio k: $(100 - k)$. As a particular case, the midpoint of P^0P^1 is given by

$$(\tfrac{1}{2}(x^1 + x^2), \tfrac{1}{2}(y^1 + y^2) + \tfrac{1}{2}(z^1 + z^2)).$$

The Distance between the point (x^0, y^0, z^0) and the line through (x^1, y^1, z^1) in direction (a, b, c):

$$\sqrt{\frac{\begin{vmatrix} y_0 - y_1 & z_0 - z_1 \\ b & c \end{vmatrix}^2 + \begin{vmatrix} z_0 - z_1 & x_0 - x_1 \\ c & a \end{vmatrix}^2 + \begin{vmatrix} x_0 - x_1 & y_0 - y_1 \\ a & b \end{vmatrix}^2}{a^2 + b^2 + c^2}}$$

Distance between the line through (x^0, y^0, z^0) in direction (a^0, b^0, c^0) and the line through (x^1, y^1, z^1)in direction (a^1, b^1, c^1):

$$\left|\frac{\begin{vmatrix} x_1 - x_0 & y_1 - y_0 & z_1 - z_0 \\ a_0 & b_0 & c_0 \\ a_1 & b_1 & c_1 \end{vmatrix}}{\sqrt{\begin{vmatrix} b_0 & c_0 \\ b_1 & c_1 \end{vmatrix}^2 + \begin{vmatrix} c_0 & a_0 \\ a_1 & b_1 \end{vmatrix}^2 + \begin{vmatrix} a_0 & b_0 \\ a_1 & b_1 \end{vmatrix}^2}}\right|$$

ANGLES

Angle between lines with directions (x^0, y^0, z^0) and (x^1, y^1, z^1): Arccos

$$\frac{a_0a_1 + b_0b_1 + c_0c_1}{\sqrt{a_0^2 + b_0^2 + c_0^2}\sqrt{a_1^2 + b_1^2 + c_1^2}}$$

In particular, the two lines are parallel when $a^0{:}b^0{:}c^0 = a^1{:}b^1{:}c1^2$, and perpendicular when $a^0\, a^1 + b^0\, b^1 + c^0\, c^1 = 0$. Angle between lines with direction angles α^0, β_0, γ_0 and α^1, β_1, γ_1: arccos(cos α^0 cos α^1 + cos β_0 cos β_1 + cos γ_0 cos γ_1)

CONCURRENCE, COPLANARITY, PARALLELISM

Two lines specified by point and direction are coplanar if and only if the determinant in the numerator of (1) is zero. In this case they are concurrent (if the denominator is nonzero) or parallel (if the denominator is zero). Three lines with directions (a^0,b^0,c^0), (a^1,b^1,c^1) and (a^2,b^2,c^2) are parallel to a common plane if and only if,

$$\begin{vmatrix} a_0 & b_0 & c_0 \\ a_1 & b_1 & c_1 \\ a_2 & b_2 & c_2 \end{vmatrix} = 0.$$

8

Geometrical Prism

INTRODUCTION

Prisms. Prisms are three-dimensional objects having end faces that are:

- Polygons;
- Parallel to each other; and
- Have the same shape and size.

Prisms have identical cross-sections if a plane cuts them parallel to the ends.

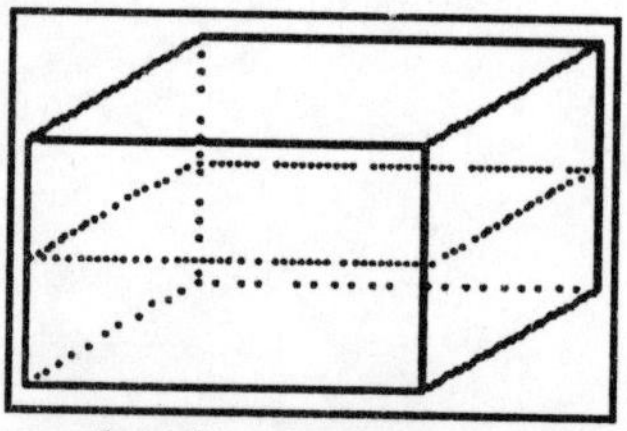

For example, a cuboid is a rectangular prism. The ends of a cuboid are rectangular and it has identical rectangular cross-sections when cut by a plane parallel to the ends. Prisms are named according to the shape of their base (or cross-section). A prism is defined as a geometric solid with two congruent bases that lie in parallel planes. You can create a prism by dragging any two-dimensional triangle or polygon through space without rotating or tilting it. The three-dimensional space defined by the moving triangle or polygon is the body of the prism. The planes at which the movement of the two-dimensional shape begins and ends are the two bases of the prism. The perpendicular distance between the parallel planes in which the bases lie is the height of the prism.

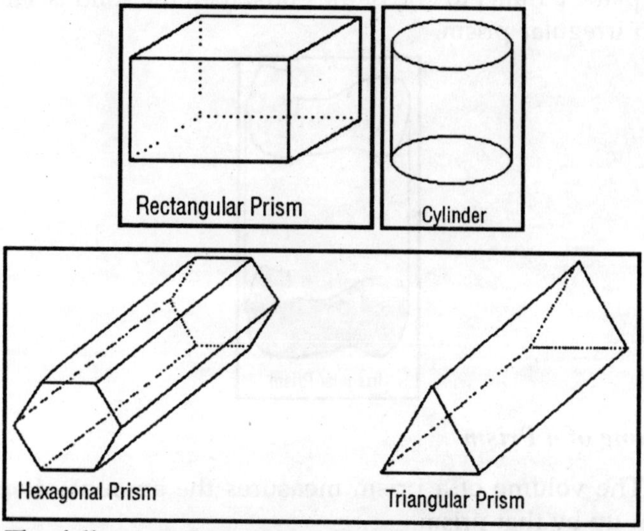

The following three-dimensional objects are not prisms, as they do not have identical cross-sections when cut by a plane parallel to the base or in the case of the sphere, parallel to a point on the sphere.

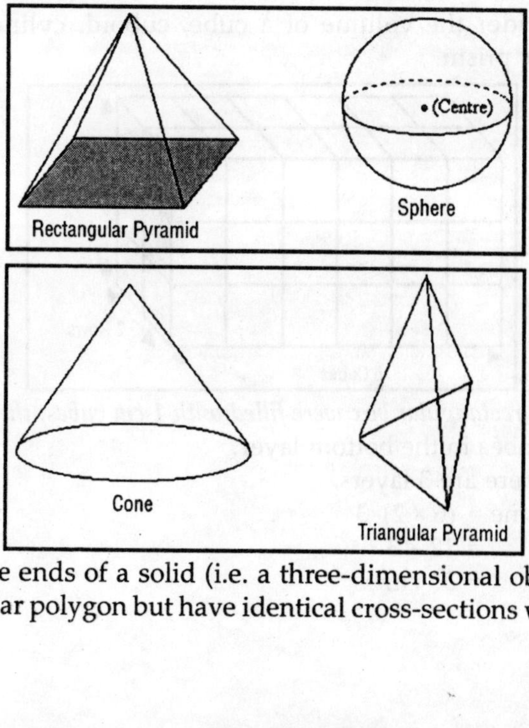

If the ends of a solid (i.e. a three-dimensional object) are not regular polygon but have identical cross-sections when cut by a plane parallel to one of the ends, then the solid is said to be an irregular prism.

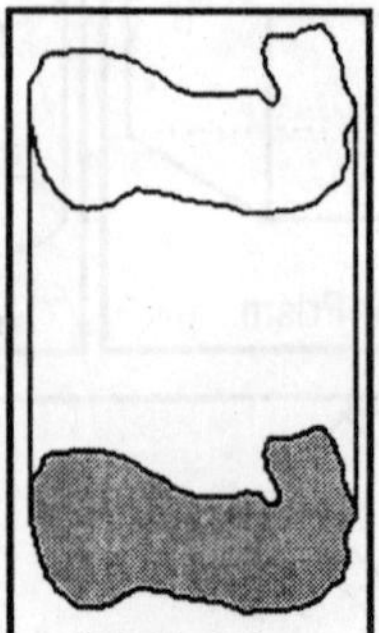

Irregular Prism

Volume of a Prism

The volume of a prism measures the amount of space taken up by that prism.

The general formula for calculating the volume of a prism is very simple:

$$\text{Volume} = Bh$$

where B is the area of the base and h is the height. In this, we will consider the volume of a cube, cuboid, cylinder and triangular prism.

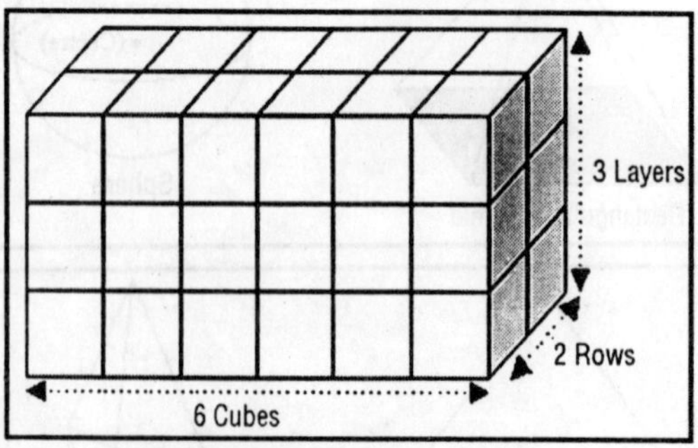

If the rectangular box were filled with 1 cm cubes, there would be: 6×2 Cubes in the bottom layer.

As there are 3 layers,

$$\begin{aligned}\text{Volume} &= (6 \times 2)\times 3\\ &= 12 \times 3\\ &= 36\ \text{cm}^3\end{aligned}$$

Now note that the area of the box's base is given by:

Area = lw

= 6 × 2

= 12 cm^2

From the above discussion, we can derive a formula for the volume of a rectangular box as follows:

Volume = 6 × 2 × 3

= (6 × 2)× 3

= Area of the base ×Height

In general: The volume, V, of a prism is given by

Volume = Area of Base × Height

V = Ah

where A is the area of the base (or cross-section) of the prism and h is the height. The volume of the following solids are often required to solve real world problems involving quantity, capacity, mass and strength of materials including liquids.

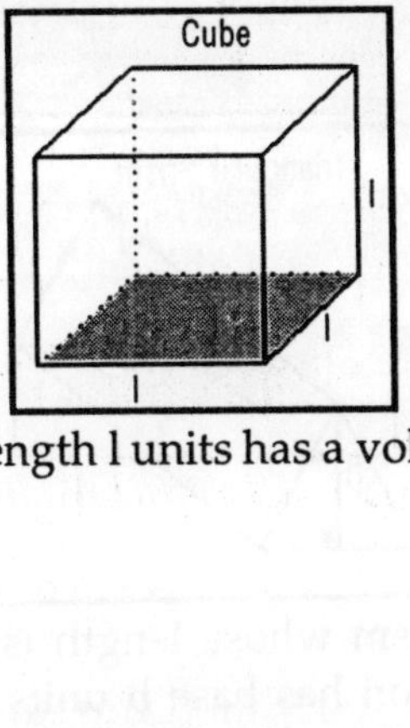

A cube of side-length l units has a volume of V cubic units given by,

$$V = Ah$$
$$= (l \times l) \times l$$
$$= l^2 \times l$$
$$= l^3$$

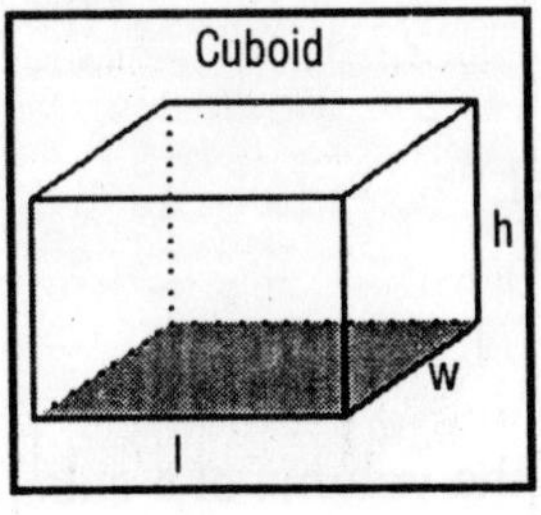

A cuboid with length l units, width w units and height h units has a volume of V cubic units given by

$$V = Ah$$
$$= (l \times w)\times h$$
$$= l \times w \times h$$
$$= lwh$$

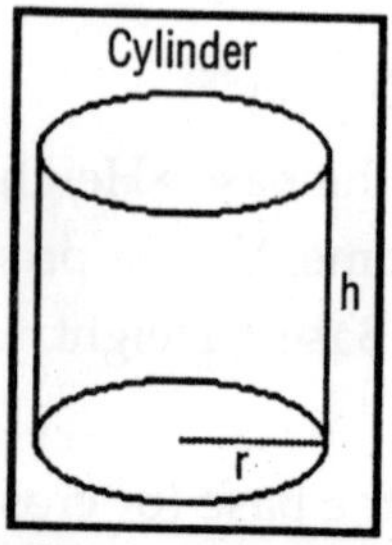

A cylinder with radius r units and height h units has a volume of V cubic units given by

$$V = Ah$$
$$= (pr^2)\times h$$
$$= p\,r^2h$$

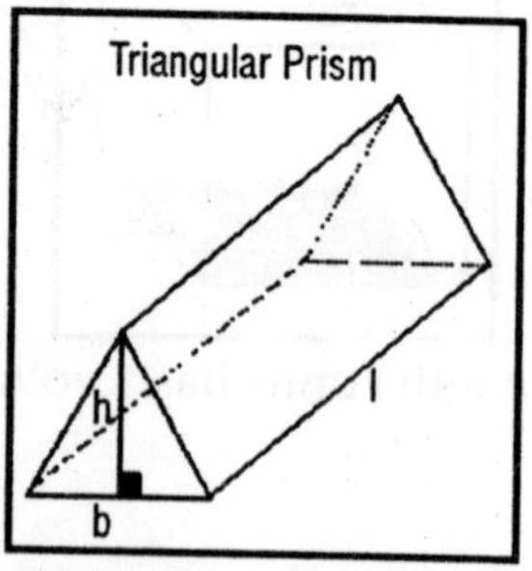

A triangular prism whose length is l units, and whose triangular cross-section has base b units and height h units, has a volume of V cubic units given by

$$V = Al$$
$$= \left(\frac{1}{2}bh\right)\times l$$
$$= \frac{1}{2}bhl$$

Example 1: Find the volume of a cube of side 5 cm.

Solution:

$V = l^3$

$= 5^3$

$= 125$

So, the Volume of the cube is 125 cm^3,

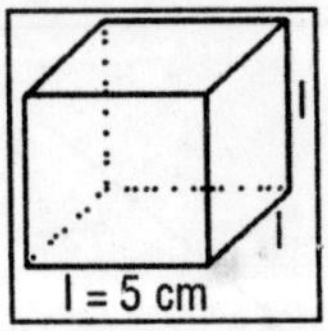

Example 2: Find the volume of a brick 10 cm by 5 cm by 4 cm.

Solution:

$V = lwh$

$= 10 \times 5 \times 4$

$= 200$

So, the volume of the brick is 200 cm^3.

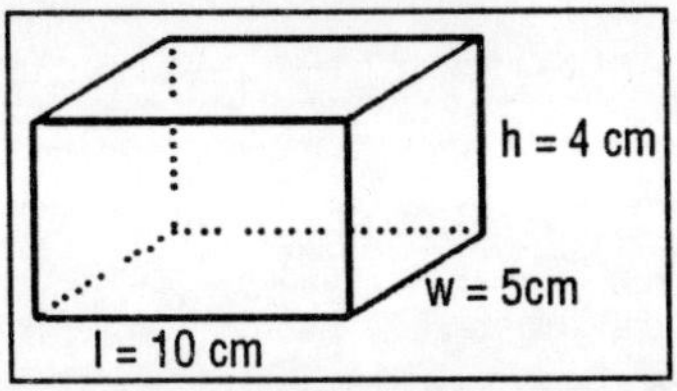

Example 3: Find the volume of the following cylinder:

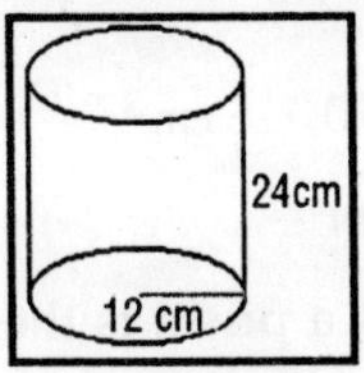

Solution:

$V = Ah$

$r = 12$ cm

$h = 24$ cm

$V = Ah$

$= pr^2h$

$= p \times 12^2 \times 24$

$= p \times 144 \times 24$

$= 10857.37$

(Correct to 2 decimal places)

So, the volume of the cylinder is 10,857.34 cm3.

Example 4: Find the volume of the following triangular prism:

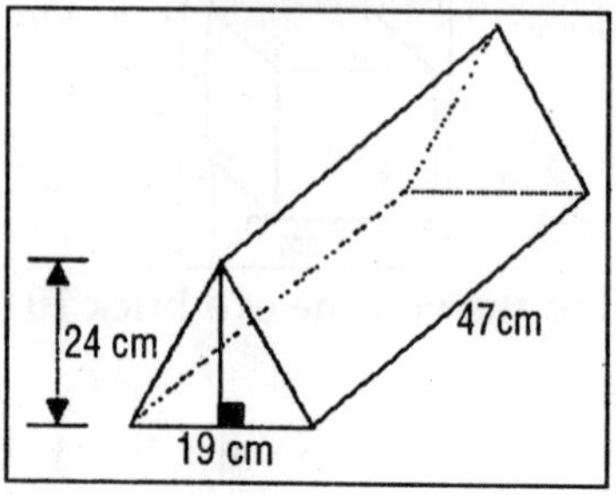

Solution:

V = Ah

b = 19 cm

h = 24

l = 47

V = Al

$$= \frac{1}{2} bhl$$

$$= \frac{1}{2} \times 19 \times 24 \times 47$$

= 107016

So, the volume is 10,716 cm^3.

Surface Area of a Prism

The surface area of a prism is the sum of the areas of all the sides of the prism. The formula for the surface area of a prism therefore depends on the type of prism you are dealing with. As with volume, we cover the specifics of calculating surface area as we cover each type of geometric solid.

RECTANGULAR SOLIDS

INTRODUCTION

We examine 3 dimensional objects that are rectangular in

shape, measuring their volume, surface area, and the length of a diagonal. In practical terms two questions about 3 dimensional objects. First, how much paint does it take to paint the object (we would have to know the surface area). Second, how much water can the object hold (we would have to know the volume). We show how to calculate surface area and volume.

A rectangular solid is a 3 dimensional object with six sides, all of which are rectangles. We first examine a cube, in which all six sides are squares. In the diagram below, a square of side 2 inches is used to form a cube with six square sides. The square clearly has an area of 4 square inches. Therefore the cube has a surface area of 48 square inches because its sides are composed of six of these squares. The cube is composed of 8 smaller cubes with a side of 1 inch. The volume of this cube is 8 cubic inches.

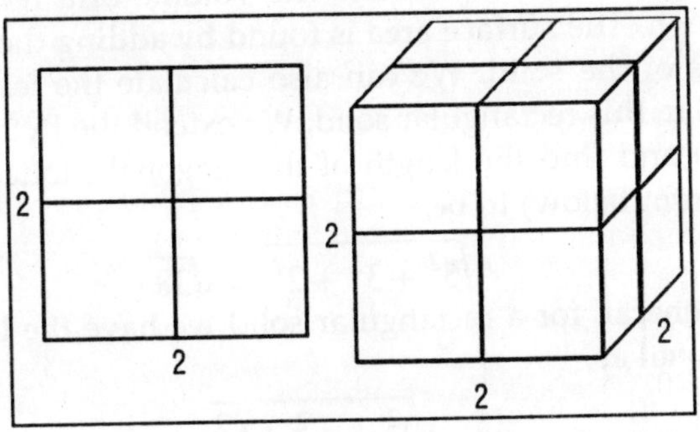

We note that a square has its name because its area is the square of the length of a side. The area $4 = 2^2$. The cube has its name because the volume is the cube of the length of a side. The volume is $8 = 2^3$. We can generalize this result for cubes by saying that the volume of a cube of side a is a^3. Since the area of one side is the length of the side squared, the entire surface area of the cube is $6 \times 4 = 24$. This can also be generalized for any cube of side a. The surface area is $6 \times a^2 = 6a^2$.

In the diagram below, we show a rectangular solid at right with dimensions $5 \times 2 \times 3$ inches. These are the measures of the length l, the width w and the height h. The area of the

Front/Back rectangles is 15 square inches. The area of the Sides is 6 square inches, and the area of the Top/Bottom rectangles is 10 square inches. Adding these we get the surface area which is 62 square inches. The volume is found by multiplying the lengths of the sides as we did with the cube. The volume is 30 cubic inches.

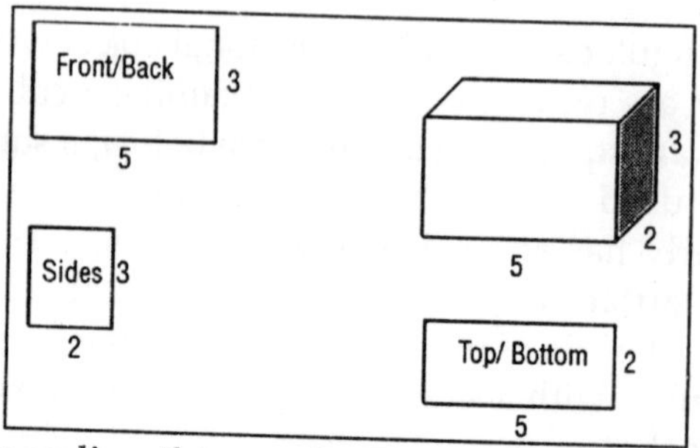

We generalize this result. The volume of a rectangular solid is lwh. The surface area is found by adding the areas of the sides of the solid. We can also calculate the length of a diagonal in this rectangular solid. We extend the Pythagorean Theorem and find the length of the diagonal (dotted line in the diagram below) to be,

$$\sqrt{5^2+3^2+2^2}=\sqrt{38}\cdot$$

In general, for a rectangular solid we have the length of the diagonal as,

$$\sqrt{l^2+w^2+h^2}\cdot$$

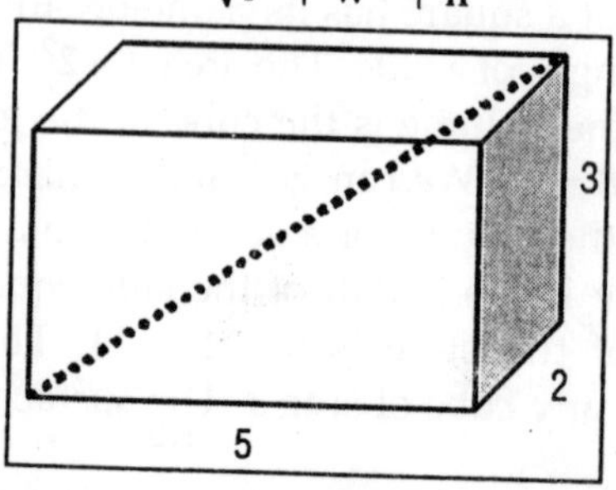

LET'S PRACTICE

- A cube has a side of 3 meters. What are the measures of the surface area, the diagonal, and the volume?

- The volume of a cube is the length of the side cubed.
- The volume is $3^3 = 27$ cubic meters. There are six sides which are squares of side 3, each having an area of 9 square meters. The total surface area is 54 square meters. The diagonal is,

$$\sqrt{3^2+3^2+3^2} = \sqrt{27} = \sqrt[3]{3}\text{ meters}\cdot$$

- A rectangular solid has dimensions 3 × 6 × 7. What are the measures of the surface area, the diagonal, and the volume? The volume is found by multiplying the measures of the sides. The volume is 126.
  There are two sides with dimensions 3 x 6 = 18, two sides with dimensions 3 × 7 = 21 and two sides with dimensions 6 × 7 = 42. The total surface area is 2(18) + 2(21) + 2(42) = 162. The diagonal is,

$$\sqrt{6^2+3^2+7^2} = \sqrt{94}\cdot$$

Volume of a Rectangular Solid

The volume of a rectangular solid is given by the following formula:

$$\text{Volume} = lwh$$

where l is the length, w is the width, and h is the height. Notice how this formula corresponds with the general formula for the volume of a prism: the product lw is the area of the base. Now, try to find the volume of the prism in the following example:

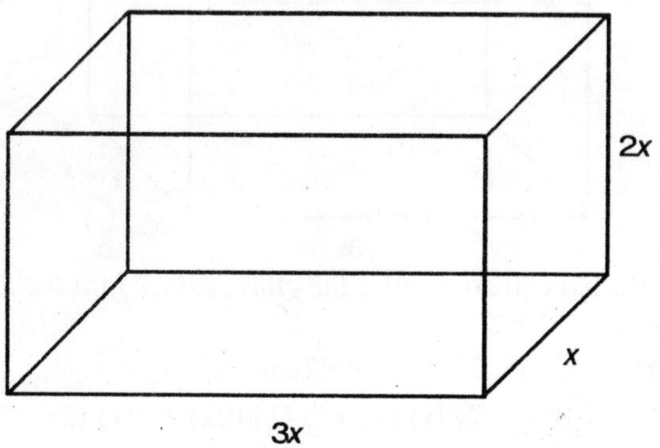

In this solid, $l = 3x$, $w = x$, and $h = 2x$. Simply plug the values into the formula given for volume, and you would find Volume = $(3x)(2x)(x) = 6x3$.

Surface Area of a Rectangular Solid

The surface area of a rectangular solid is given by the following formula:

$$\text{Surface Area} = 2/w + 2/h + 2wh$$

where l is the length, w is the width, and h is the height.

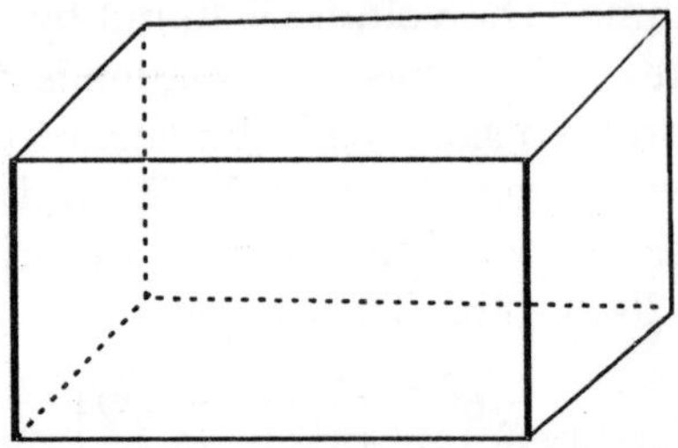

The six faces of a rectangular solid consist of three congruent pairs. The surface area formula is derived by simply adding the areas of the faces-two faces have areas of $l \times w$, two faces have areas of $l\ h$, and two faces have areas of $w \times$ h. To practice, try to find the surface area of the rectangular solid we used as an example for volume.

Here's the figure again:

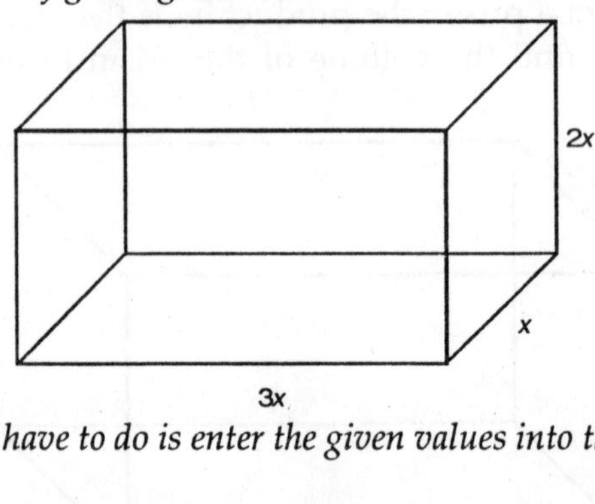

All you have to do is enter the given values into the formula for surface area:

$$\begin{aligned}\text{Surface Area} &= 2lw + 2lh + 2wh \\ &= 2(3x)\,(x) + 2(3x)\,(2x) + 2(x)\,(2x) \\ &= 6x^2 + 12x^2 + 4x^2 = 22x^2\end{aligned}$$

Diagonal Length of a Rectangular Solid

The diagonal of a rectangular solid, *d*, is the line segment whose endpoints are opposite corners of the solid. Every rectangular solid has four diagonals, each with the same length, that connect each pair of opposite vertices. Here's one diagonal drawn in:

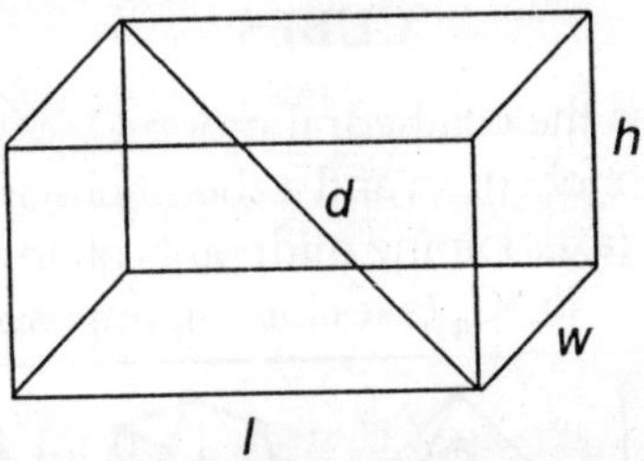

The formula for the length of a diagonal is:

$$d = \sqrt{l^2 + w^2 + h^2}$$

where l is the length, w is the width, and h is the height.

You can look at this formula as the Pythagorean theorem in three dimensions. In fact, you can derive this formula using the Pythagorean theorem. First, find the length of the diagonal along the base.

This is $\sqrt{l^2 + w^2}$. Then use the Pythagorean theorem again, incorporating height to find the length of the diagonal from one corner to the other:

$$d^2 = (\sqrt{l^2 + w^2}) + h^2.$$

Thus, $d^2 = l^2 + w^2 + h^2$ and $d = \sqrt{l^2 + w^2 + h^2}$. A Math IIC question might ask you: What is the length of diagonal *AH* in the rectangular solid below if *AC* = 5, *GH* = 6, and *CG* = 3?

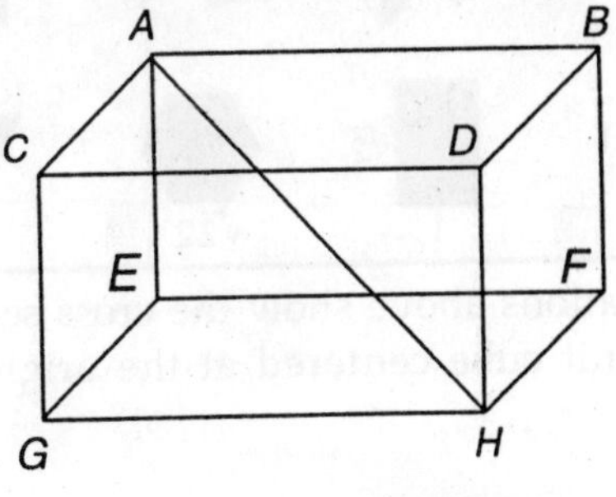

The question gives the length, width, and height of the rectangular solid, so you can just plug those numbers into the formula:

$$AH = \sqrt{5^2 + 6^2 + 3^2} = \sqrt{25 + 36 + 9} = \sqrt{70}$$

The length of the diagonal AH (as well as BG, CF, and DE) is $\sqrt{70}$.

CUBES

The cube has the octahedral group O_h of symmetries, and is an equilateral zonohedron and a rhombohedron. It has 13 axes of symmetry: $6C_2$ (axes joining midpoints of opposite edges), $4C_3$ (space diagonals), and $3C_4$ (axes joining opposite face centroids).

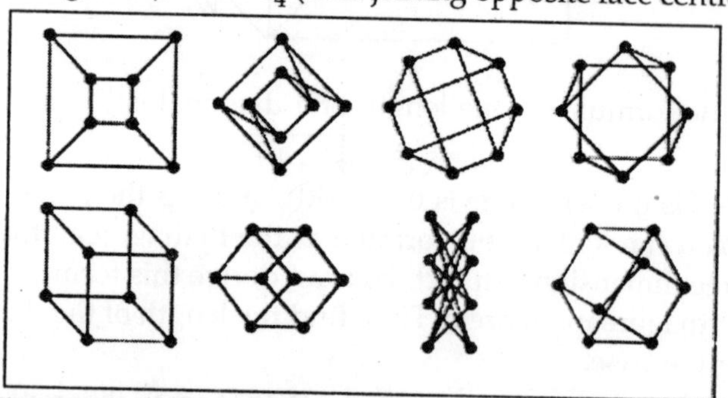

The connectivity of the vertices of the cube is given by the cubical graph. Using so-called "wallet hinges," a ring of six cubes can be rotated continuously.

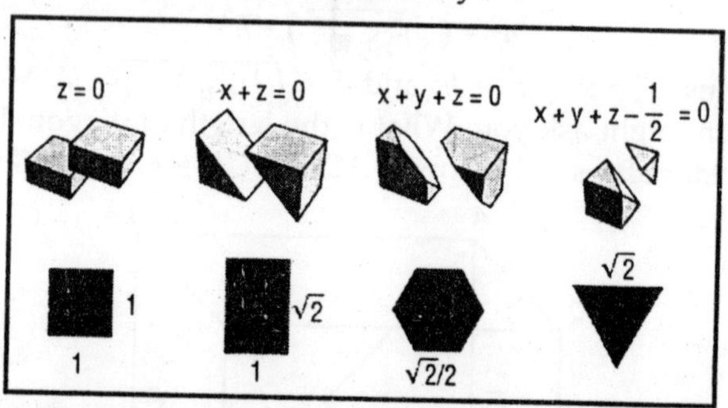

The illustrations above show the cross sections obtained by cutting a unit cube centered at the origin with various planes.

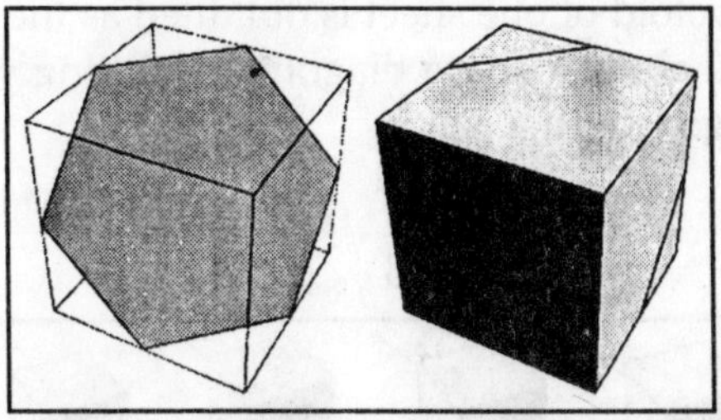

As shown above, a plane passing through the midpoints of opposite edges (perpendicular to a C_3axis) cuts the cube in a regular hexagonal cross section. Since there are four such axes, there are four possible hexagonal cross sections. If the vertices of the cube are $(\pm1,\pm1,\pm1)$, then the vertices of the inscribed hexagon are (0, –1,–1),(1,0,–1),(1,1,0), (0,1,1), (–1,0,1), and (–1,–1,0). A hexagon is also obtained when the cube is viewed from above a corner along the extension of a space diagonal.

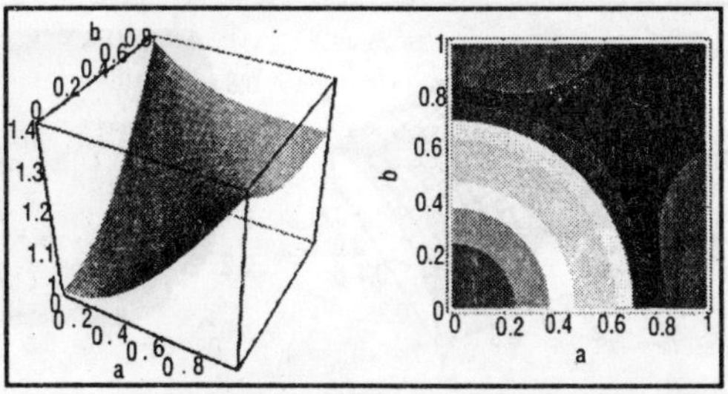

The maximal cross sectional area that can be obtained by cutting a unit cube with a plane passing through its center is $\sqrt{2}$, corresponding to a rectangular section intersecting the cube in two diagonally opposite edges and along two opposite face diagonals. The area obtained as a function of normal to the plane (a, b, 1) is illustrated above.

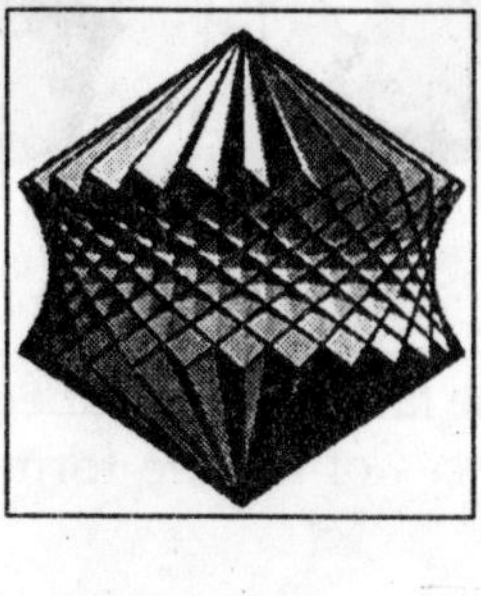

A hyperboloid of one sheet is obtained as the envelope of a cube rotated about a space diagonal resulting volume for a cube with edge length a is,

$$V = \frac{1}{3}\sqrt{3}\pi a^3$$

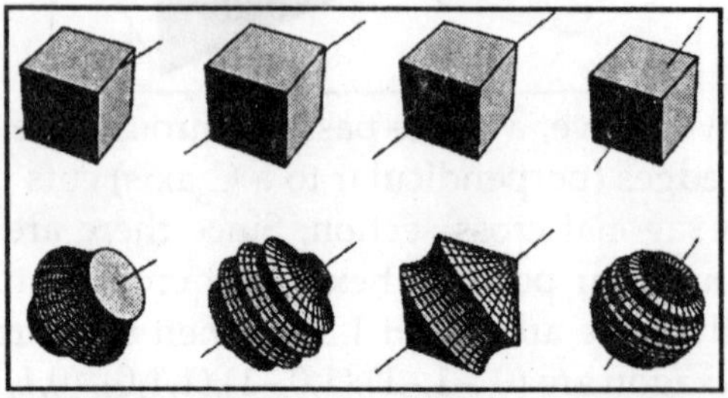

More generally, consider the solid of revolution obtained for revolution axis passing through the center and the point (x, y,l), several examples of which are shown above.

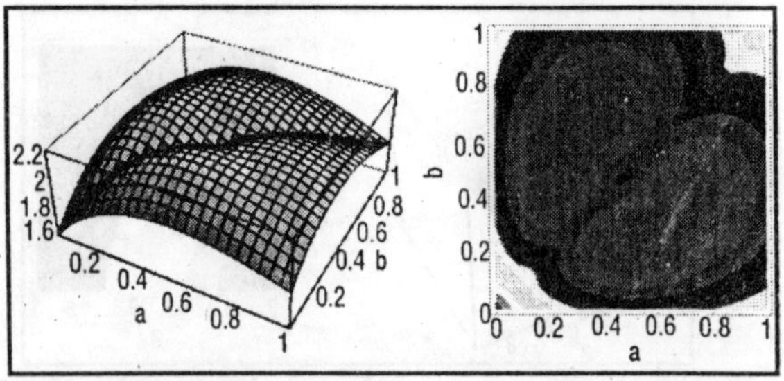

As shown by Hidekazu, the solid with maximum volume is obtained for parameters of approximately (a, b) = (0.529307, 0.237593). This corresponds to the rightmost plot above.

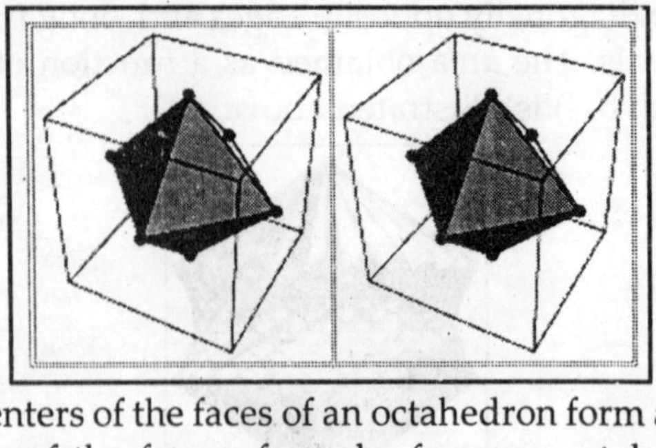

The centers of the faces of an octahedron form a cube, and the centers of the faces of a cube form an octahedron. The largest square which will fit inside a cube of edge length a has each corner a distance 1/4 from a corner of a cube. The resulting square has edge length $3\sqrt{2}\,a/4$, and the cube containing that edge is called Prince Rupert's cube.

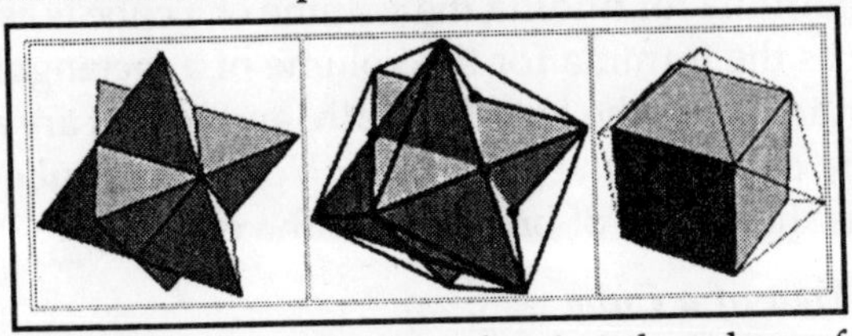

The solid formed by the faces having the edges of the stella octangula (left figure) as polygon diagonals is a cube. Affixing a square pyramid of height 1/2 on each face of a cube having unit edge length results in a rhombic dodecahedron.

Since its eight faces are mutually perpendicular or parallel, the cube cannot be stellated. The cube can be constructed by cumulation of a unit edge-length tetrahedron by a pyramid with height,

$$\frac{1}{6}\sqrt{6}.$$

The following table gives polyhedra which can be constructed by cumulation of a cube by pyramids of given heights h.

h	(r + h)/h	Result
$\frac{1}{6}$	4/3	Tetrakis Hexahedron
$\frac{1}{2}$	2	Rhombic Dodecahedron
$\frac{1}{2}\sqrt{2}$	$1+\sqrt{2}$	24-Faced Star Deltahedron

The polyhedron vertices of a cube of edge length 2 with face-centered axes are given by $(\pm1,\pm1,\pm1)$. If the cube is oriented with a space diagonal along the z-axis, the coordinates are, $(0, \sqrt{3})$, $(0, 2\sqrt{2/3}, 1/\sqrt{3})$, $(\sqrt{2}, \sqrt{2/3}, -1/\sqrt{3})$, $(\sqrt{2}, -\sqrt{2/3}, 1/\sqrt{3})$, $(0, -2\sqrt{2/3}, -1\sqrt{3})$, $(-\sqrt{2}, -\sqrt{2/3}, 1/\sqrt{3})$, $(-\sqrt{2}, -\sqrt{2/3}, -1/\sqrt{3})$

and the negatives of these vectors. A faceted version is the great cubicuboctahedron.

Volume of a Cube

The formula for finding the volume of a cube is essentially the same as the formula for the volume of a rectangular solid. However, in a cube the length, width, and height are all equal, so that the cube volume formula is: Volume of Cube = s^3 where s is the length of one edge of the cube.

Surface Area of a Cube

Since a cube is just a rectangular solid whose sides are all equal, the formula for finding the surface area of a cube is the same as for a rectangular solid, except with $s = l = w = h$: Surface Area of a Cube = $6s^2$ where s is the length of one edge of the cube.

Diagonal Length of a Cube

The same goes for the diagonal length of a cube. The formula for the diagonal of a cube is simply adapted from the formula for the diagonal length of a rectangular solid, with $s = l = w = h$: $d = \sqrt{3s^2} = s\sqrt{3}$ where s is the length of one edge of the cube.

CYLINDERS

A cylinder is a prism with a circular base. The height of a cylinder, h, is the length of the line segment whose endpoints are the centers of the bases. The radius of a cylinder, r, is the radius of its base. If you know the height and radius of a cylinder, you can calculate its volume and surface area quite easily.

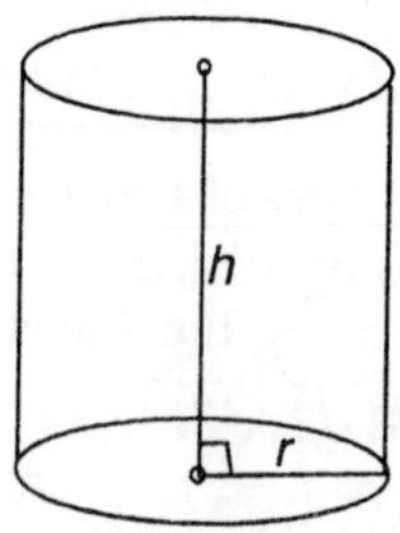

Volume of a Cylinder

The volume of a cylinder is the product of the area of its base with its height. Because a cylinder has a circular base, the volume of a cylinder is: Volume of a Cylinder = $\pi r^2 h$ where r is the radius of the circular base and h is the height. Try to find the volume of the cylinder below.

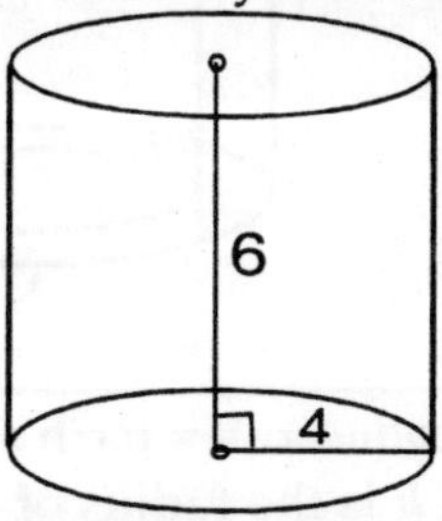

This cylinder has a radius of 4 and a height equal to 6.

Using the formulas:

$$\text{Volume} = \pi (4)^2 (6) = 96\pi$$

A cylinder is a solid that has two parallel faces which are congruent circles. These faces form the bases of the cylinder. The cylinder has one curved surface. The height of the cylinder is the perpendicular distance between the two bases.

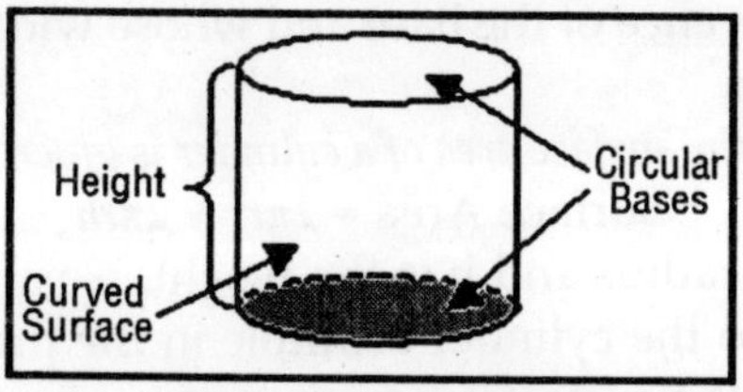

The volume of a cylinder is given by the formula:

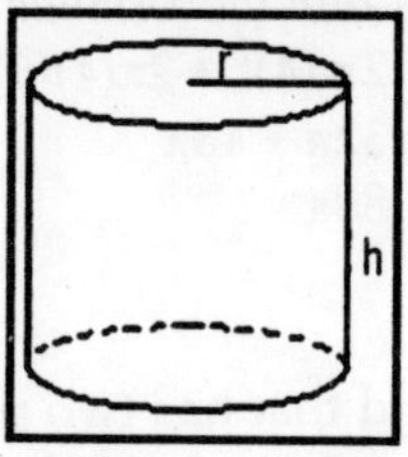

- Volume = Area of base × height
- V = pr^{2}h where r = radius of cylinder and *h* is the height or length of cylinder.

Volume of hollow cylinder: Sometimes you may be required to calculate the volume of a hollow cylinder or tube.

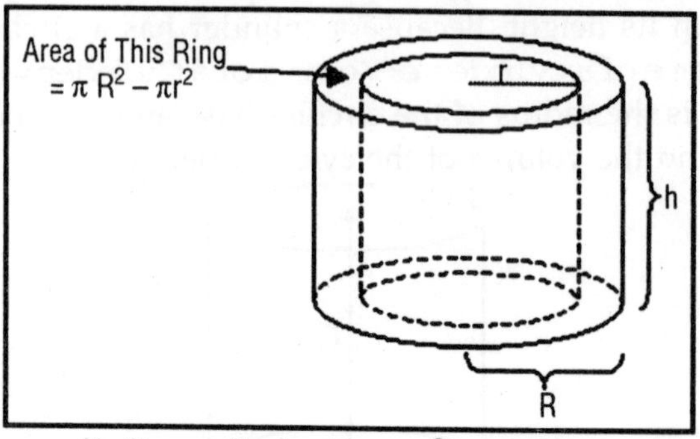

Volume of hollow cylinder: V = p r^2h where *r* is the radius of the outer surface and *h* is the radius of the inner surface.

Surface Area of a Cylinder

The surface area of a cylinder is the sum of the areas of the two bases and the lateral face of the cylinder. The bases are congruent circles, so their areas can be found easily. The lateral face is the tubing that connects the two bases. When "unrolled," the lateral base is simply a rectangle whose length is the circumference of the base and whose width is the height of the cylinder.

Therefore, the surface area of a cylinder is given by this formula:

$$\text{Surface Area} = 2\pi r^2 + 2\pi rh$$

where *r* is the radius and h is the height.

Go back to the cylinder example in the previous section, and now find the surface area.

Plugging the values into the formula:

$$\begin{aligned}\text{Surface Area} &= 2\pi\,(4)^2 + 2\pi(4)(6)\\ &= 32\pi + 48\pi\\ &= 80\pi\end{aligned}$$

Description

A cylinder is a solid that has two parallel faces which are congruent circles. These faces form the bases of the cylinder. The cylinder has one curved surface. The height of the cylinder is the perpendicular distance between the two bases.

The net of a solid cylinder consists of 2 circles and one rectangle. The curved surface opens up to form a rectangle.

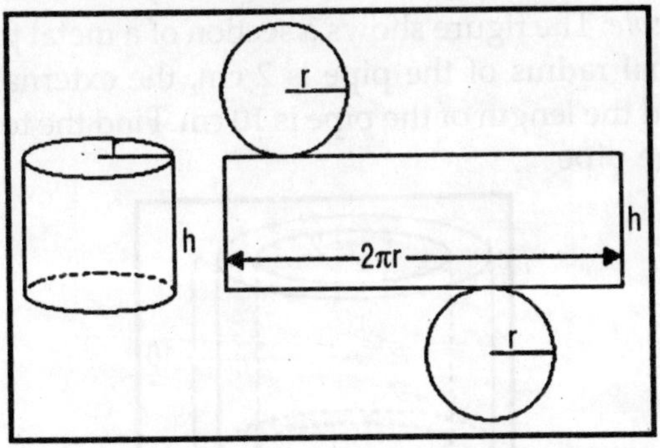

Surface area = 2 × area of circle + area of rectangle,

$$A = 2\times(\pi r^2)(2\pi r\times h) = 2\pi r^2 + 2\pi rh$$

Example: The diameter of the base of a cylinder is 12 cm and the height is 8 cm. Find the surface area of the solid cylinder.

Solution: Radius = 6 cm

Surface area = 2pr (r + h)

$$= 2\times\frac{22}{7}\times 6\times(6+8)$$

= 528 cm2

Surface Area of Hollow Cylinder

Sometimes you may be required to calculate the total surface area of a hollow cylinder or tube.

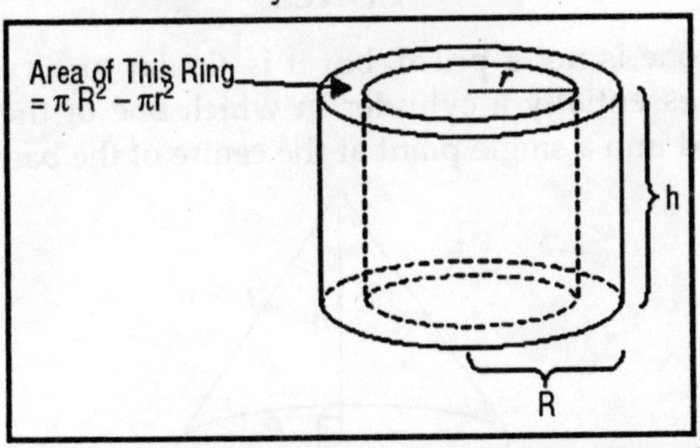

Total surface area of hollow cylinder = area of internal

curved surface + area of external curved surface + area of the two rings $= 2\pi rh + 2\pi Rh + 2(\pi R^2 - \pi r^2)$

Example: The figure shows a section of a metal pipe. Given the internal radius of the pipe is 2 cm, the external radius is 2.4 cm and the length of the pipe is 10 cm. Find the total surface area of the pipe

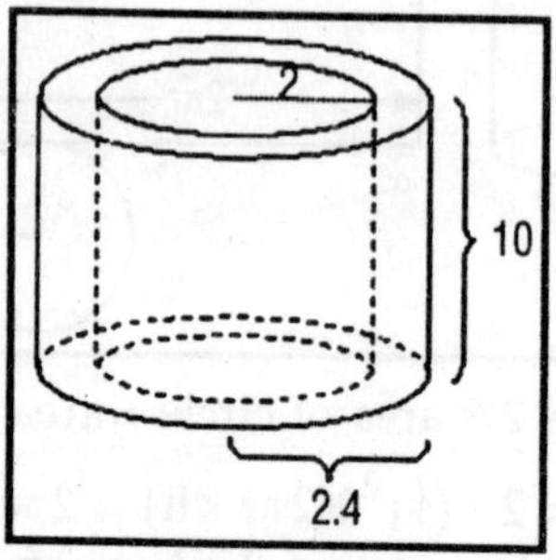

Solution: r = 2, R = 2.4, h = 10

Total surface area of pipe = area of internal surface + area of external surface + area of the two rings,

= 2prh + 2pRh + 2(pR^2- pr^2)

= (2p × 2 × 10) + (2p × 2.4 × 10) + (2 × (2.4^{2}p - 2^{2}p))

= 40p + 48p + 3.52p

= 91.52p

= 91.52 × 3.142

= 287.56 cm^2

CONES

A cone is not a prism, but it is similar to a cylinder. A cone is essentially a cylinder in which one of the bases is collapsed into a single point at the centre of the base.

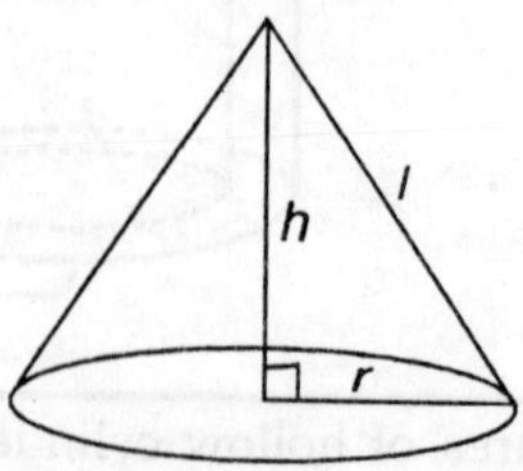

The radius of a cone is the radius of its one circular base. The height of a cone is the distance from the centre of the base to the apex (the point on top). The lateral height, or slant height, of a cone is the distance from a point on the edge of the base to the apex. In the previous figure, these three measurements are denoted by *r*, *h*, and *l*, respectively.

Notice also that the height, radius, and lateral height of a cone make up a right triangle. This means that if you know the value for any two of these measurements, you will always be able to find the third by using the Pythagorean theorem.

Volume of a Cone

Since a cone is similar to a cylinder, except collapsed to a single point at one end, the formula for the volume of a cone is a fraction of the formula for the volume of a cylinder:

A cone is a solid with a circular base. It has a curved surface which tapers (i.e. decreases in size) to a vertex at the top. The height of the cone is the perpendicular distance from the base to the vertex.

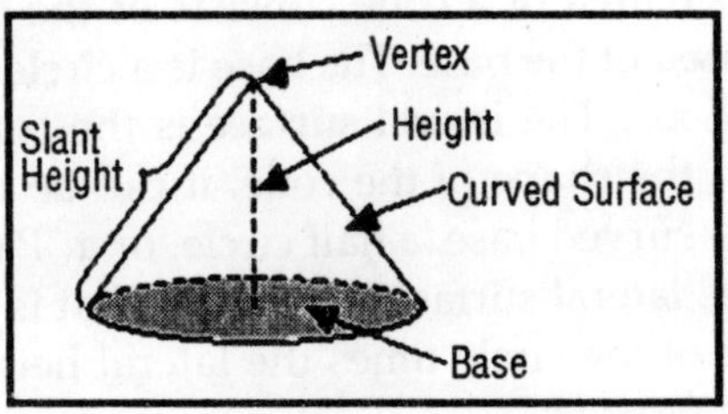

The volume of a cone is given by the formula:

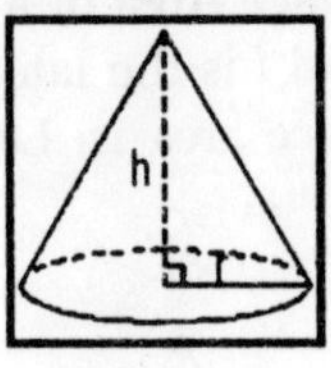

$$\text{Volume of cone} = \frac{1}{3} \times \text{Area of base} \times \text{height},$$

$$V = V = \frac{1}{3}\pi r^2 h$$

where *r* is the radius of the base and *h* is the height of the prism.

Volume of Coue $= \frac{1}{3}$ Volume of a Cylinder $= \frac{1}{3}\pi r^2 h$ where r is the radius and h is the height.

For practice, find the volume of the cone pictured below:

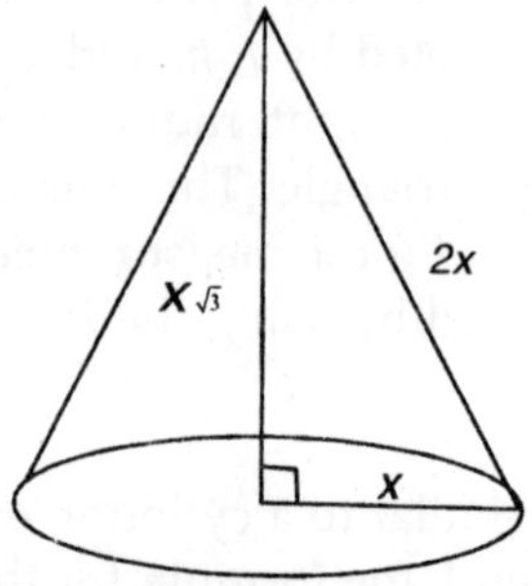

To answer this question, just use the formula for the volume of a cone with the following values plugged in: $r = x$, $l = 2x$, and $h = x\sqrt{3}$. The volume is $1/3\ \pi(x^2)(x\sqrt{3}) = x\ \pi/3$.

Surface Area of a Cone

The surface area of a cone consists of the lateral surface area and the area of the base. The base is a circle and therefore has an area of πr^2. The lateral surface is the cone "unrolled." Depending on the shape of the cone, it can be the shape of a triangle with a curved base, a half circle, or a "Pacman" shape. The area of the lateral surface is a length that is related to the circumference of the circle times the lateral height, l.

This is the formula:

Lateral Surface Area of a Coue $= \pi rl$

where r is the radius and l is the lateral height. An alternate formula for lateral surface area is: Lateral Surface Area of a Coue $= \pi \frac{1}{3} cl$

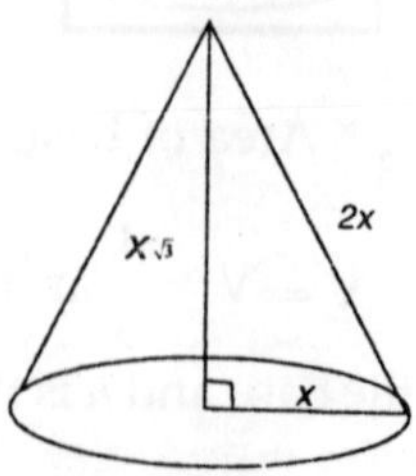

where *c* is the circumference and *l* is still the lateral height. The two formulas for surface area are the same, since the circumference of a circle is equal to $2\pi r$. But knowing one or the other can save time.

The total surface area of a cone is the sum of the base area and lateral surface area:

$$\text{Total Surface Area of a Cane} = \pi r^2 + \pi rl$$

When you are finding the surface area of a cone, be careful not to just find the lateral surface area and then stop. Students often neglect to add the area of the circular base. Practice by finding the total surface area of the cone pictured below:

The total surface area is equal to the area of the base plus the lateral surface area. The area of the base = πx^2. The lateral surface area = $\pi x \times 2x$. The total surface area therefore equals $\pi x^2 + \pi 2x^2 = 3\pi x^2$.

Example: Calculate the volume of a cone if the height is 12 cm and the radius is 7 cm.

Solution: Volume

$$= \frac{1}{3} \times \pi r^2 h$$

$$= \frac{1}{3} \times \frac{22}{7} \times 7^2 \times 12$$

$$= 616\text{cm}^3$$

Surface Area of Cone: A cone is a solid with a circular base. It has a curved surface which tapers (i.e. decreases in size) to a vertex at the top. The height of the cone is the perpendicular distance from the base to the vertex.

The net of a solid cone consists of a small circle and a sector of a larger circle. The arc of the sector has the same length as the circumference of the smaller circle.

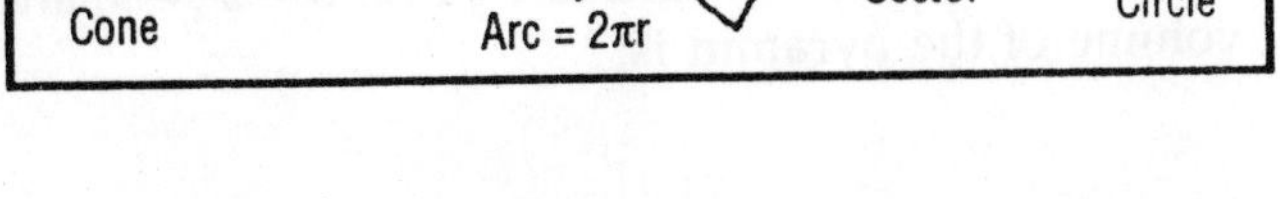

Surface area of cone = Area of sector + area of circle,

$$= \pi rs + \pi r^2 = \pi r(r+s)$$

Example: A cone has a circular base of radius 10 cm and a slant height of 30 cm. Calculate the surface area.

Solution:

$$\text{Area} = \pi r(r + s)$$

$$= \frac{22}{7} \times 10 \times (10+30)$$

$$= 1{,}257.14 \text{ cm2}$$

PYRAMIDS

A pyramid is like a cone, except that it has a polygon for a base. Though pyramids are not tested very often on the Math IIC test, you should be able to recognize them and calculate their volume.

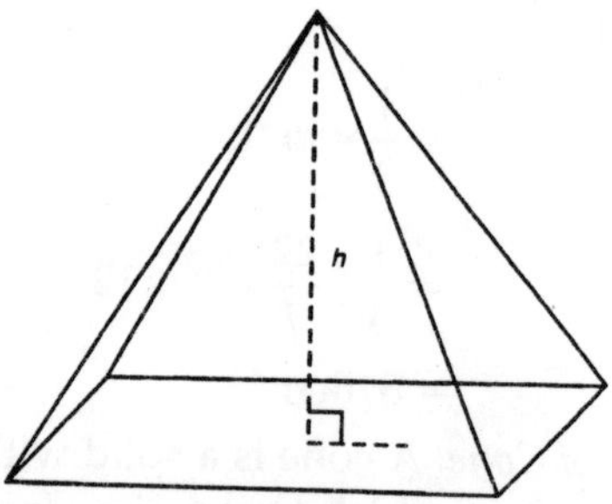

The shaded area in the figure above is the base, and the height is the perpendicular distance from the apex of the pyramid to its base.

Volume of a Pyramid

The formula for calculating the volume of a pyramid is:

$$\text{Volume} = \frac{1}{3} Bh$$

where *B* is the area of the base and *h* is the height. Try to find the volume of the pyramid below: The base is just a square with a side of 3, and the height is $3\sqrt{2}/2$. $B = 3^2 = 9$, and the total volume of the pyramid is:

$$\text{Volume} = \frac{1}{3} \times (9)\left(\frac{3\sqrt{2}}{2}\right)$$

$$= \frac{1}{3} \times \frac{27}{\sqrt{2}}$$

$$= \frac{9\sqrt{2}}{2}$$

A pyramid is a solid with a polygonal base and several triangular lateral faces. The pyramid is named after the shape of its base. For example, rectangular pyramid, triangular pyramid. The lateral faces meet at a common vertex. The height of the pyramid is the perpendicular distance from the base to the vertex.

Example:

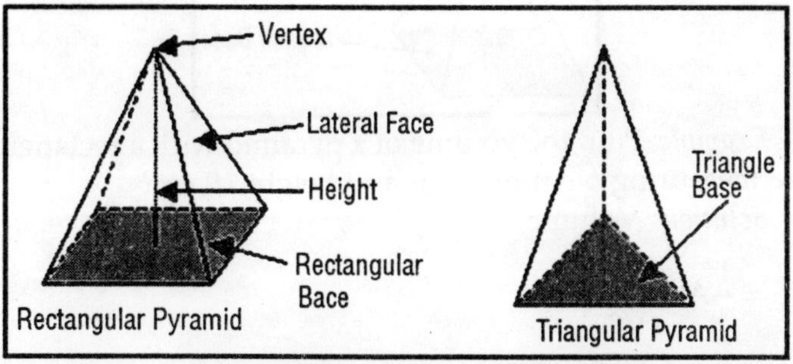

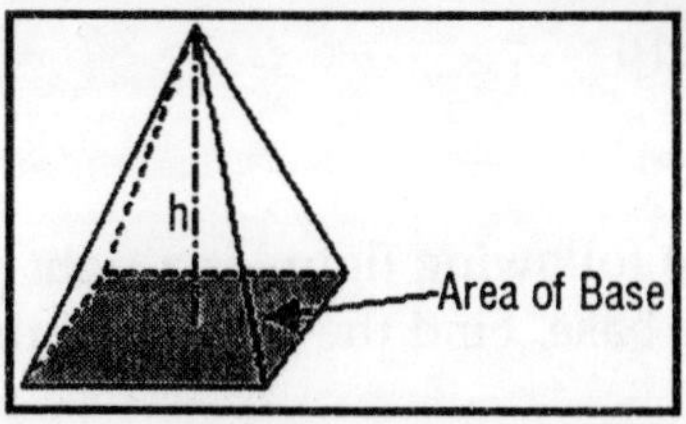

The volume of a pyramid is given by the formula: Volume of pyramid = $\frac{1}{3}\times$ Area of base × height $V = \frac{1}{3}Ah$ where A is the area of the base and h is the height of the pyramid.

Surface Area of a Pyramid

The surface area of a pyramid is rarely tested on the Math IIC test. If you do come across a question that covers the topic, you can calculate the area of each face individually using techniques from plane geometry, since the base of the pyramid is a square and the sides are triangles.

Example: The volume of a pyramid is given by the formula:

Volume

$$= \frac{1}{3} \times \text{Area of base} \times \text{height}$$

$$= \frac{1}{3} Ah$$

where A is the area of the base and h is the height of the pyramid.

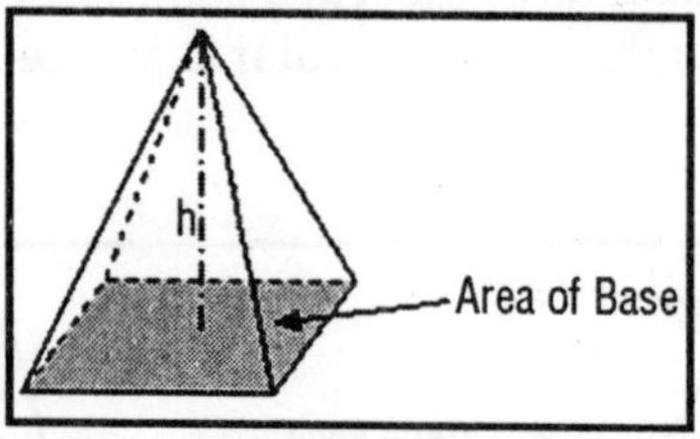

Example: Find the volume of a pyramid with a rectangular base measuring 6 cm by 4 cm and height 10 cm.

Solution: Volume

$$= \frac{1}{3} Ah$$

$$= \frac{1}{3} \times 6 \times 4 \times 10$$

= 80 cm3

Example: The following figure is a right pyramid with an isosceles triangle base. Find the volume of the pyramid if the height is 20 cm.

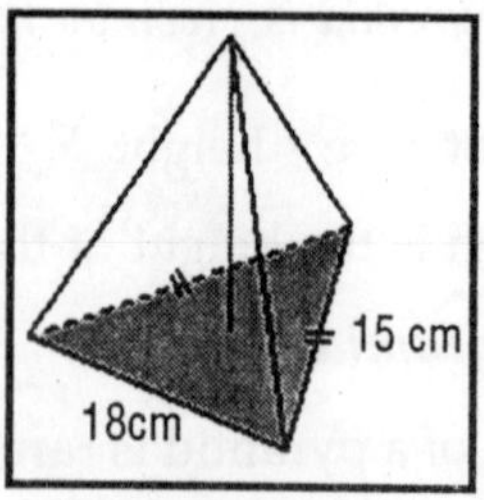

Solution: First, we have to calculate the area of the base.

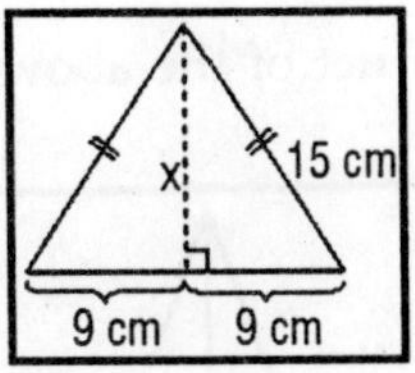

To do that, we would need to get the height of the isosceles triangle that forms the base. Using Pythagoras' theorem,

$$x = \sqrt{15^2 - 9^2}$$

$= 12$ cm

Area of triangle:

$$= \frac{1}{2} \times 18 \times 12$$

$= 108$ cm^2

Volume of pyramid:

$$\frac{1}{3}Ah$$

$$\frac{1}{3} \times 108 \times 20cm^3$$

$= 720$ cm^3

Surface area of Pyramid: A pyramid is a solid with a polygonal base and several triangular lateral faces. The pyramid is named after the shape of its base. The lateral faces meet at a common vertex. The height of the pyramid is the perpendicular distance from the base to the vertex.

Example: Calculate the surface area of the following pyramid.

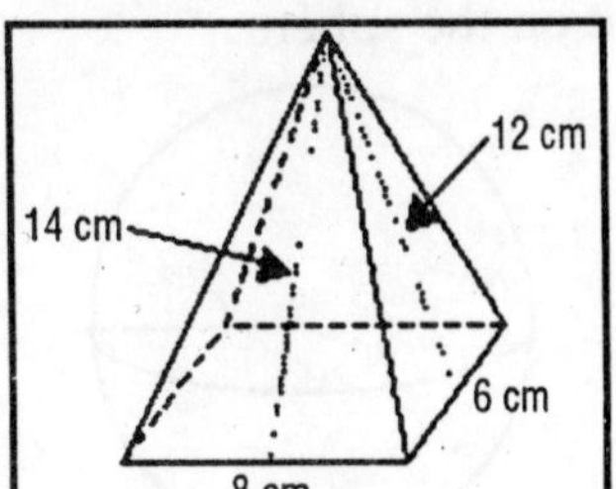

Solution: Sketch a net of the above pyramid to visualize the surfaces.

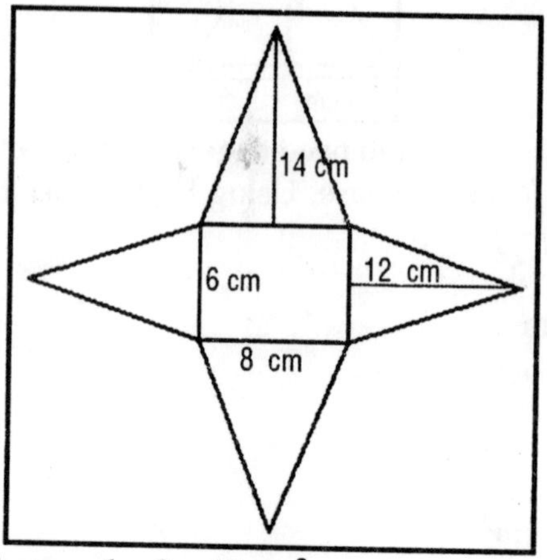

Area of base = $6 \times 8 = 48\ cm^2$

Area of two of the triangles = $\frac{1}{2} \times 2 \times 6 \times 12 = 72\ cm^2$

Area of other two triangles = $\frac{1}{2} \times 2 \times 8 \times 14 = 112\ cm^2$

Total surface area = $48 + 72 + 112 = 232\ cm^2$

SPHERES

DEFINITION

A sphere is the collection of points in three-dimensional space that are equidistant from a fixed point, the centre of the sphere. Essentially, a sphere is a 3-D circle. The main measurement of a sphere is its radius, r, the distance from the centre to any point on the sphere.

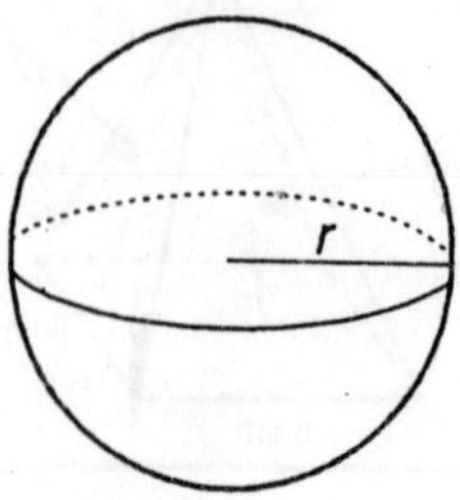

If you know the radius of a sphere, you can figure out both its volume and surface area.

The equation for the volume of a sphere is:

$$\text{Volume of a Sphere} = \frac{4\pi r^2}{3}$$

The equation for the surface area of a sphere is: Surface Area of Sphere $= 4\pi r^2$

DESCRIPTION

A sphere is a solid in which all the points on the round surface are equidistant from a fixed point, known as the centre of the sphere. The distance from the centre to the surface is the radius.

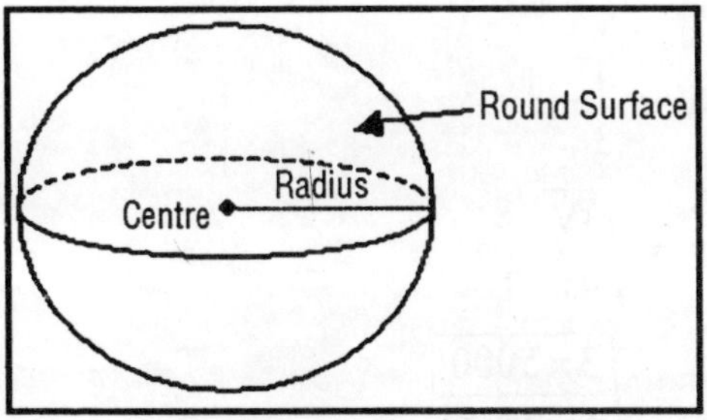

Volume of sphere = $\frac{4}{3}\pi r^3$ where r is the radius.

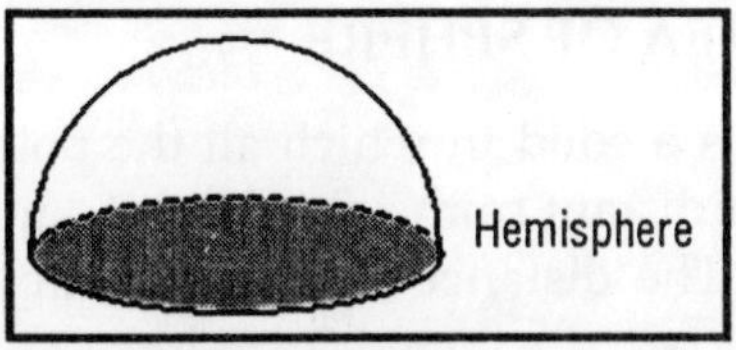

A hemisphere is half a sphere, with one flat circular face and one bowl-shaped face.

Volume of hemisphere,

$$= \frac{1}{2}\left(\frac{4}{3}\pi r^3\right) = \frac{2}{3}\pi r^3$$

where *r* is the radius.

Example: Calculate the volume of sphere with radius 4 cm.

Solution: Volume of sphere,

$$\frac{4}{3}\pi r^3$$

$$\frac{4}{3}\times\frac{22}{7}\times 4^3$$

$$= 268.19\text{cm}$$

We can also change the subject of the formula to obtain the radius given the volume.

$$V = \frac{4}{3}\pi r^3$$

Example: The volume of a spherical ball is 5,000 cm 3. What is the radius of the ball?

Solution:

$$V = \frac{4}{3}\pi r^3$$

$$r = \sqrt{\frac{3V}{4\pi}}$$

$$= \sqrt{\frac{3\times 5000}{4\times 3.142}}$$

$$= 10.61\text{cm}$$

SURFACE AREA OF SPHERE

A sphere is a solid in which all the points on the round surface are equidistant from a fixed point, known as the centre of the sphere. The distance from the centre to the surface is the radius.

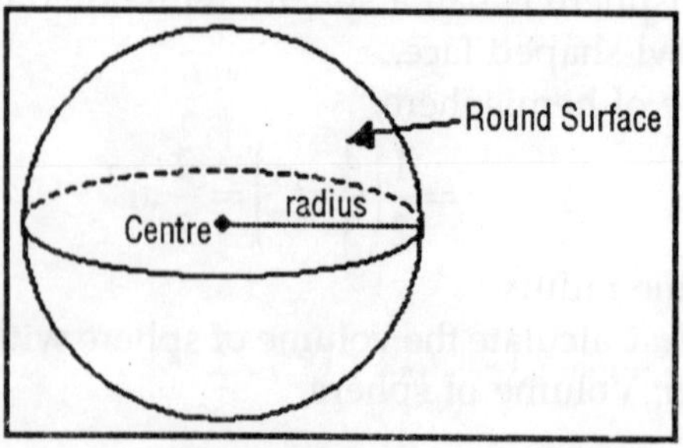

Surface area of a sphere with radius r = $4\pi r^2$

Example: Calculate the surface area of a sphere with radius 3.2 cm

Solution: Surface area of sphere

= 4p r^2

= 4p $(3.2)^2$

= 4 × 3.14 × 3.2 × 3.2

= 128.6 cm^2

Example: Calculate the surface area of a hemisphere with radius 4 cm

Solution: Surface area of hemisphere

$= \frac{1}{2} \times$ surface area of sphere + area of base

$= \frac{1}{2} \times 4p\, r^2 + p$

= 3p r^2

DIMENSIONS

The first way the Math IIC will test your understanding of the relationship between the basic measurements of geometric solids is by giving you the length, surface area, or volume of various figures and asking you to compare their dimensions. The math needed to answer comparing dimensions questions isn't all that hard. But in order to get to the math, you need to have a good grasp of the formulas for each type of solid and be able to relate those formulas to each other algebraically.

For example: The surface area of a sphere is the same as the volume of a cylinder. What is the ratio of the radius of the sphere to the radius of the cylinder?

This question tells you that the surface area of a sphere and the volume, a cylinder are equal. A sphere's surface area is $4\pi(r_s)^2$, where r_s is the radius of the sphere.

A cylinder's volume is $\pi(r_c)^2 \times h$, where r_c is the radius of the cylinder and h is its height.

Therefore:

$$4\pi\,(r_s)^2 = \pi\,(r_c)^2 \times h$$

The question asks for the ratio between the radii of the sphere and cylinder. This ratio is given by r_s/r_c. Now you can solve the equation $4\pi r_s^2 = \pi r_c^2 \times h$ for the ratio r_s/r_c.

$$4\pi (r_s)^2 = \pi (r_c)^2 \times h$$

$$4(r_s)^2 = h (r_c)^2$$

$$\frac{(r_s)^2}{(r_c)^2} = \frac{h}{4}$$

$$\frac{r_s}{r_c} = \sqrt{\frac{h}{4}}$$

MEASUREMENTS

When the solids in the question are increased by a single constant factor, a simple rule can help you come to an answer.

- If a solid's length is multiplied by a given factor, then its surface area is multiplied by the square of that factor, and its volume is multiplied by the cube of that factor.

Remember that this rule holds true if *all* of a solid's dimensions increase in length by a given factor. So for a cube or a sphere, just a side or the radius need change for the rule to hold, but for a rectangular solid, cylinder, or other solid, all of the dimensions must change. If the dimensions of the object do not increase by a constant factor, for instance, if the height of a cylinder doubles but the radius of the base triples, you will have to go back to the equation for the dimension you are trying to determine and calculate by hand.

Example: If you double the length of the side of a square, by how much do you increase the area of that square?

If you understand the formula for the area of a square, this question is not difficult. The formula for the area of a square is $A = s^2$, where s is the length of a side. Replace s with $2s$, and you see that the area of a square quadruples when the length of its sides double: $(2s)^2 = 4s^2$.

Example: If a sphere's radius is halved, by what factor does its volume decrease?

The radius of the sphere is multiplied by a factor of 1/2 (or divided by a factor of 2), and so its volume multiplies by the cube of that factor: $(1/2)^3 = 1/8$. Therefore, the volume of the sphere is multiplied by a factor of 1/8 (divided by 8), which is the same as decreasing by a factor of 8.

Example: A rectangular solid has dimensions $x \times y \times z$ (these are its length, width, and height) and has a volume of 64. What is the volume of a rectangular solid of dimensions $x/2 \times y/2 \times z$? If this rectangular solid had dimensions that were all one-half as large as the dimensions of the solid whose volume is 64, then its volume would be $(1/2)^3 \times 64 = 1/8 \times 64 = 8$. But dimension z is not multiplied by 1/2 like x and y. Hence, to answer a question like this one, you should turn to the volume formula for rectangular solids: volume = $l \times w \times h$. It is given in the question that $xyz = 64$. So, $x/2 \times y/2 \times z = 1/4\ xyz = 1/4 \times 64 = 16$.

INSCRIBED SOLIDS

An inscribed solid is a solid fit inside another solid, with the edges of the two solids touching. The figures below are, from left to right, a cylinder inscribed in a sphere, a sphere inscribed in a cube, and a rectangular solid inscribed in a sphere.

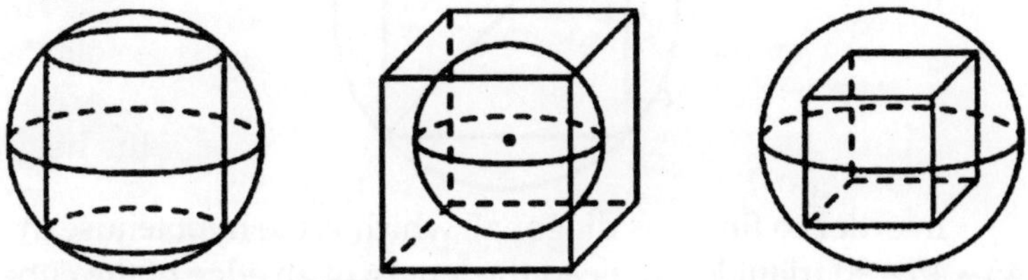

Most often, an inscribed solid question will present a figure of an inscribed solid and give you information about one of the solids. For example, you may be given the radius of a cylinder and then be asked to find the volume of the other solid using the figure as your guide. Use what you know about the radius of the cylinder to find the dimensions of the other solid. Take a look at this example: In the figure below, a cube is inscribed in a cylinder. If the length of the diagonal of the cube is 4 and the height of the cylinder is 5, what is the volume of the cylinder?

The formula for the volume of a cylinder is $\pi r^2(h)$. It is given in the question that $h = 5$, but there is no value given for

r. So in order to solve for the volume of the cylinder, we need to first find the value of r. The key step in this problem is to recognize that the diagonal of a face of the cube is also the diameter, or twice the radius, of the cylinder.

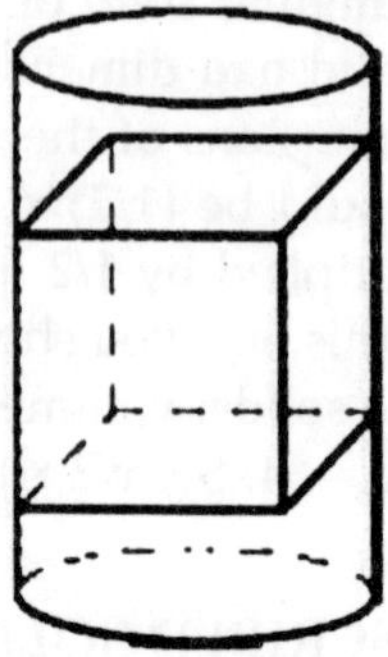

To see this, draw a diagonal, d, in either the top or bottom face of the cube.

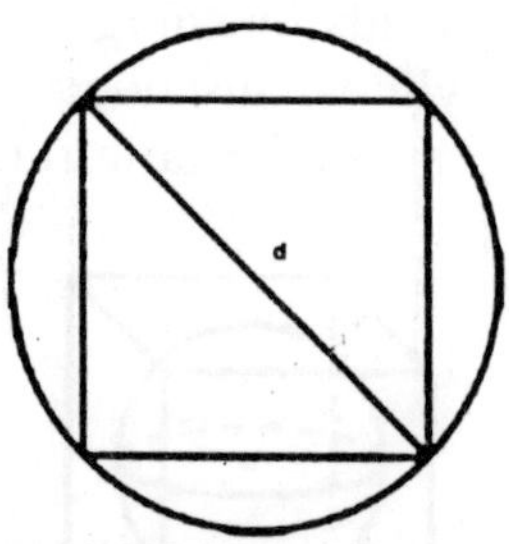

In order to find this diagonal, which is the hypotenuse in a 45 – 45 – 90 triangle, we need the length of an edge of the cube, or s. We can find s from the diagonal of the cube (not to be mistaken with the diagonal of a face of the cube), since the formula for the diagonal of a cube is $s\sqrt{3}$ with s being the length of an edge of the cube. The question gives us the diagonal of the cube as $4\sqrt{3}$ so it follows that $s = 4$. This means that the diagonal along a single face of the cube is $4\sqrt{2}$ (using the special properties of a 45 – 45 – 90 triangle). Therefore, the radius of the cylinder is $4\sqrt{2}/2 = 2\sqrt{2}$. Plug that back into the formula for the volume of the cylinder, and you get $\pi \times (2\sqrt{2})^2 \times 5 = 40\pi$.

RULES OF INSCRIBED SOLIDS

Math IIC questions involving inscribed solids are much

easier to solve when you know how the lines of different solids relate to one another. For instance, the previous example showed that when a cube is inscribed in a cylinder, the diagonal of a face of the cube is equal to the diameter of the cylinder. The better you know the rules of inscribed solids, the better you'll do on these questions.

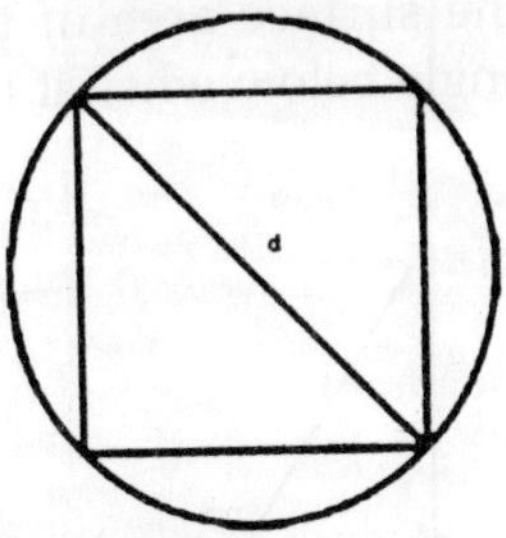

Fig. Cylinder Inscribed in a Sphere

The diameter of the sphere is equal to the diagonal of the cylinder's height and diameter.

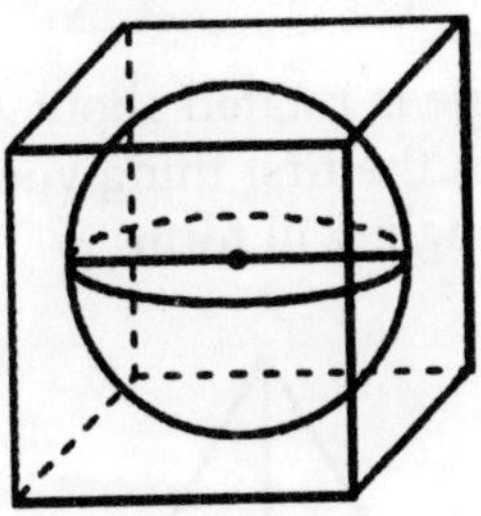

Fig. Sphere Inscribed in a Cube

The diameter of the sphere is equal to the length of cube's edge.

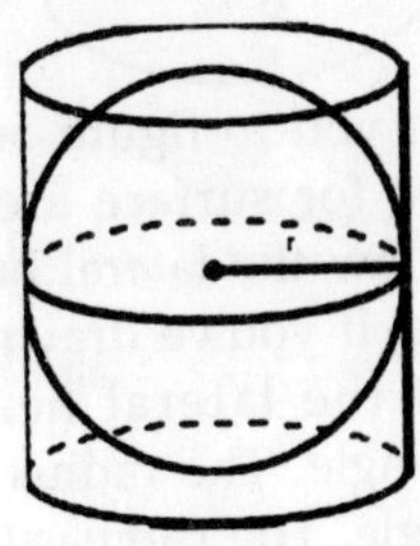

Fig. Sphere Inscribed in a Cylinder

The cylinder and the sphere have the same radius.

POLYGON ROTATION

another type of Math IIC question that you may come across involves a solid produced by the rotation of a polygon. The best way to explain this type of problem is with an example: What is the surface area of the geometric solid produced by the triangle below when it is rotated 360° about the axis *AB*?

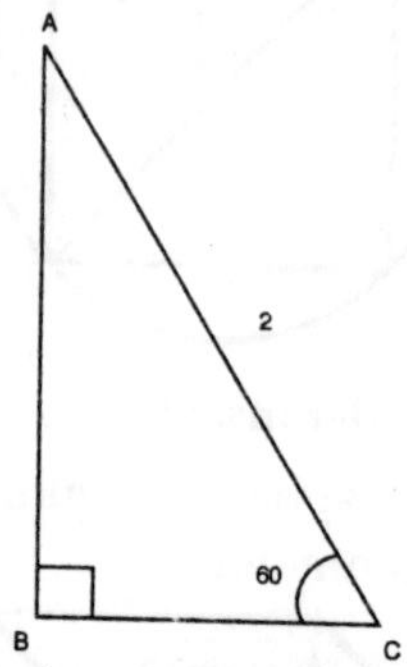

When this triangle is rotated about *AB*, a cone is formed. To solve the problem, the first thing you should do is sketch the cone that the triangle will form.

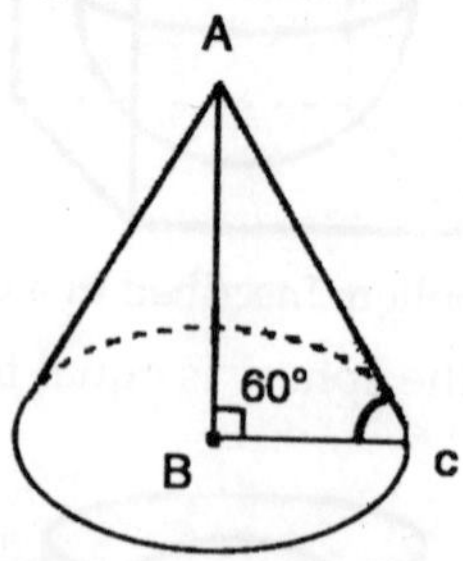

The question asks you to figure out the surface area of the cone. The formula for surface area is $\pi r^2 + \pi rl$, which means you need to know the lateral height of the cone and the radius of the circle. If you've drawn your cube correctly, you should see that the lateral height is equal to the hypotenuse of the triangle. The radius of the circle is equal to side *BC* of the triangle. You can easily calculate the length of *BC* since the triangle is a 30 – 60 – 90 triangle. If the hypotenuse is 2, then *BC*, being the side opposite the 30º angle, must be 1.

Now plug both values of l and r into the surface area formula and then simplify:

$$\text{Total Surface Area} = \pi(1)^2 + \pi(1)(2) = \pi + 2\pi = 3\pi$$

COMMON ROTATIONS

You don't need to learn any new techniques or formulas for problems dealing with rotating figures. You just have to be able visualize the described rotation and be aware of which parts of the polygons become which parts of the geometric solid. Below is a summary of which polygons, when rotated a specific way, produce which solids. A rectangle rotated about its edge produces a cylinder.

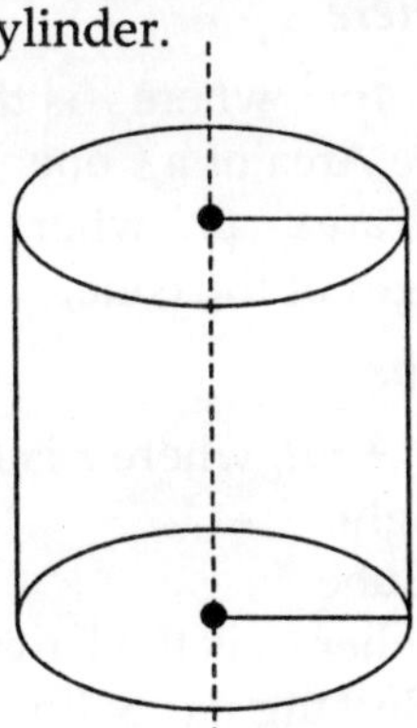

A semicircle rotated about its diameter produces a sphere.

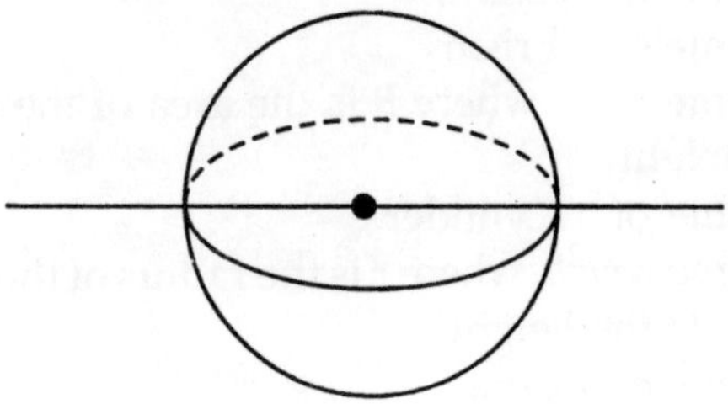

FORMULAS

Length of a Diagonal of a Cube $d = \sqrt{3s^2} = s\sqrt{3}$. where *s* is the length of one edge of the cube. Length of a Diagonal of a Rectangular Solid,

$$d = \sqrt{l^2 + w^2 + h^2},$$

where l is the length, w is the width, and h is the height of the rectangular solid.

Surface Area of a Cube

- Surface area = $6s^2$, where s is the length of one edge.
- *Surface Area of a Rectangular Solid*: Surface area = $2lw + 2lh + 2wh$, where l is the length, w is the width, and h is the height.

Surface Area of a Cylinder

Surface area = $2\pi r^2 + 2\pi rh$, where r is the radius and h is the height.

Surface Area of a Sphere

- Surface area = $4pr^2$, where r is the radius.
- Lateral Surface Area of a Cone
- Lateral surface area = prl, where r is the radius and l is the lateral height of the cone.

Surface Area of a Cone

Surface area = $\pi r^2 + \pi rl$, where r is the radius of the base and l is the lateral height.

- Volume of a Cube
- Volume = s^3, where s is the length of one edge.
- Volume of a Rectangular Solid
- Volume = lwh, where l is the length, w is the width, and h is the height.
- Volume of a Prism
- Volume = Bh, where B is the area of the base and h is the height.
- Volume of a Cylinder
- Volume = $\pi r^2 h$, where r is the radius of the circular base and h is the height.
- Volume of a Cone
- Volume = $\frac{1}{3}\pi r^2 h$, where r is the radius and h is the height.

- Volume of a Pyramid
- Volume $=\frac{1}{3}$ *Bh*, where *B* is the area of the base and *h* is the height.
- Volume of a Sphere
- Volume $=\frac{4\pi r^3}{3}$, where *r* is the radius.

9

Differential Polyhedra

DESCRIPTION

For any polyhedron topologically equivalent to a sphere in particular, for any convex polyhedron the Euler formula holds: $v - e + f = 2$, where v is the number of vertices, e is the number of edges, and f is the number of faces.

Many common polyhedra are particular cases of cylinders or cones. A cylinder with a polygonal base (directrix) is called a prism. A cone with a polygonal base is called a pyramid. A frustum of a cone with a polygonal base is called a trucated pyramid.

A prism whose base is a parallelogram is a parallelepiped. The volume of a parallelepiped with one vertex at the origin and adjacent vertices at (x^1,y^1,z^1), (x^2,y^2,z^2) and (x^3,y^3,z^3) is given by,

$$\frac{1}{6}\begin{vmatrix} x_1 & y_1 & z_1 \\ x_2 & y_2 & z_2 \\ x_3 & y_3 & z_3 \end{vmatrix}$$

The rectangular parallelepiped is a particular case: all its faces are rectangles. If the side lengths are a, b, c, the volume is abc, the total area is $2(ab + ac + bc)$, and each diagonal has length $\sqrt{a^2 + b^2 + c^2}$.

When $a = b = c$ we get the cube: A pyramid whose base is a triangle is a tetrahedron. The volume of a tetrahedon with one vertex at the origin and the other vertices at (x^1,y^1,z^1), (x^2,y^2,z^2) and (x^3,y^3,z^3) is given by,

$$\frac{1}{6}\begin{vmatrix} x_1 & y_1 & z_1 \\ x_2 & y_2 & z_2 \\ x_3 & y_3 & z_3 \end{vmatrix}$$

In a tetrahedron with vertices P^0, P^1, P^2, P^3, let d_{ij} be the distance (edge length) from P^i to P_j. Form the determinants

$$\Delta = \begin{vmatrix} 0 & 1 & 1 & 1 & 1 \\ 1 & 0 & d_{01}^2 & d_{02}^2 & d_{03}^2 \\ 1 & d_{01}^2 & 0 & d_{12}^2 & d_{13}^2 \\ 1 & d_{02}^2 & d_{12}^2 & 0 & d_{23}^2 \\ 1 & d_{03}^2 & d_{13}^2 & d_{23}^2 & 0 \end{vmatrix} \text{ and } \Gamma = \begin{vmatrix} 0 & d_{01}^2 & d_{02}^2 & d_{03}^2 \\ d_{01}^2 & 0 & d_{12}^2 & d_{13}^2 \\ d_{02}^2 & d_{12}^2 & 0 & d_{23}^2 \\ d_{03}^2 & d_{13}^2 & d_{23}^2 & 0 \end{vmatrix}$$

Then the volume of the tetrahedron is,

$$\sqrt{|\Delta|/288},$$

and the radius of the circumscribed sphere is,

$$\sqrt{\frac{1}{2}|\Gamma/\Delta|}.$$

REGULAR POLYHEDRA

Figure shows the five regular polyhedra, or Platonic solids.

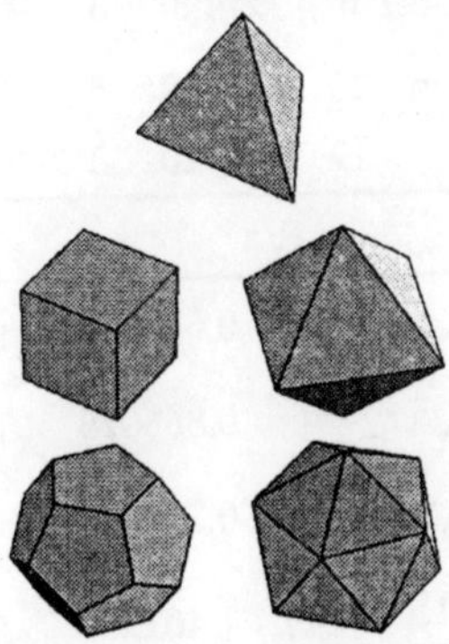

Fig. *The Platonic solids. Top:* The Tetrahedron (self-dual). Middle: the cube and the Octahedron (dual to one another). Bottom: the Dodecahedron and the Icosahedron (Dual to one another).

In the following tables and formulas, *a* is the length of an edge, θ the dihedral angle at each edge, R the radius of the circumscribed sphere, r the radius of the inscribed sphere, V the volume, S the total area, v the total number of vertices, e the total number of edges, f the total number of faces, p the number of edges in a face, and q the number of edges meeting at a vertex.

$$q = 2 \arcsin \frac{\cos(180^\circ/q)}{\sin(180^\circ/p)}$$

$$\frac{R}{a} = \frac{\frac{1}{2}\sin(180^\circ/q)}{\sin(180^\circ/p)\cos\frac{1}{2}\theta}$$

$$\frac{R}{r} = \tan\frac{180^\circ}{p}\tan\frac{180^\circ}{q}$$

$$\frac{S}{a^2} = \frac{fp}{4}\cot\frac{180^\circ}{p}$$

$$V = \frac{1}{3}rS$$

Name	v	e	f	p	q	sin θ	θ
Regular tetrahedron	4	6	4	3	3	$2\sqrt{2}/3$	70°31′44″
Cube	8	12	6	4	3	1	90°
Regular octahedron	6	12	8	3	4	$2\sqrt{2}/3$	109°28′16″
Regular dodecahedron	20	30	12	5	3	$2\sqrt{5}$	116°33′54″
Regular icosahedron	12	30	20	3	5	2/3	138°11′23″

Name	R/a		r/a	
Tetrahedron	$\sqrt{6}/4$	0.612372	$\sqrt{6}/12$	0.204124
Cube	$\sqrt{3}/2$	0.866025	$\frac{1}{2}$	0.5
Octahedron	$\sqrt{2}/2$	0.707107	$\sqrt{6}/6$	0.408248
Dodecahedron	$\frac{1}{4}\left(\sqrt{15}+\sqrt{3}\right)$	1.401259	$\frac{1}{20}\sqrt{250+110\sqrt{5}}$	1.1113516
Icosahedron	$\frac{1}{4}\sqrt{10+2\sqrt{5}}$	0.951057	$\frac{1}{12}\sqrt{42+18\sqrt{5}}$	0.755761

Name	S/a^2		V/a^3	
Tetrahedron	$\sqrt{3}$	0.73205	$\sqrt{2}/12$	0.117851
Cube	6	6	–1	1–
Octahedron	$2\sqrt{3}$	3.46410	$\sqrt{2}/3$	0.471405
Dodecahedron	$3\sqrt{25+10\sqrt{5}}$	20.64573	$\frac{1}{4}\sqrt{15+7\sqrt{5}}$	7.663119
Icosahedron	$5\sqrt{3}$	8.66025	$\frac{5}{12}(3+\sqrt{5})$	2.181695

CYLINDERS

Given a line L and a curve C in a plane P, the cylinder with generator L and directrix C is the surface obtained by moving L parallel to itself, so that a point of L is always on C.

If L is parallel to the z-axis, the surface's implicit equation does not involve the variable z. Conversely, any implicit equation that does not involve one of the variables (or that can be brought to that form by a change of coordinates) represents a cylinder. If C is a simple closed curve, we also apply the word cylinder to the solid enclosed by the surface generated in this way.

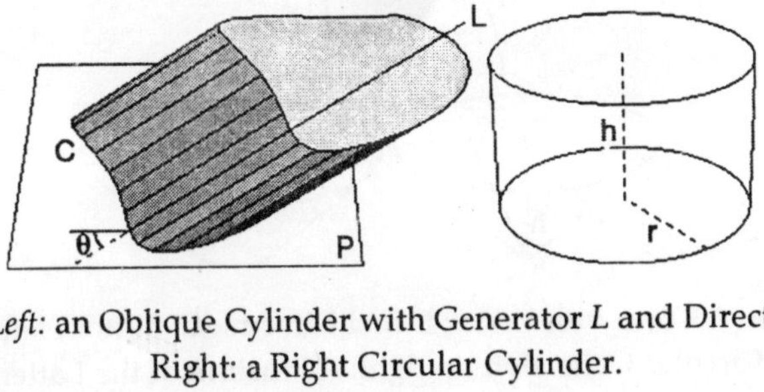

Fig. *Left:* an Oblique Cylinder with Generator L and Directrix C. Right: a Right Circular Cylinder.

The volume contained between P and a plane P' parallel to P is,

$$V = Ah = Al \sin \theta,$$

where A is the area in the plane P enclosed by C, h is the distance between P and P' (measured perpendicularly), l is the length of the segment of L contained between P and P', and θ is the angle that L makes with P. When $\theta = 90°$ we have a right cylinder, and $h = 1$. For a right cylinder, the lateral area between P and P' is hs, where s is the length (circumference) of C.

The most important particular case is the right circular cylinder (often simply called a cylinder). If r is the radius of the base and h is the altitude, the lateral area is, $2\pi rh$, the total area is, $2\pi r(r + h)$, and the volume is $\pi r^2 h$. The implicit equation of this surface can be written $x^2 + y^2 = r^2$.

CONES

Given a curve C in a plane P and a point O not in P, the cone with vertex O and directrix C is the surface obtained as the union of all lines that join O with points of C. If O is the origin and the surface is given implicity by an algebraic equation, that equation is homogeneous (all terms have the same total degree in the variables). Conversely, any homogeneous implicit equation (or one that can be made homogeneous by a change of coordinates) represents a cone. If C is a simple closed curve, we also apply the word cone to the solid enclosed by the surface generated in this way.

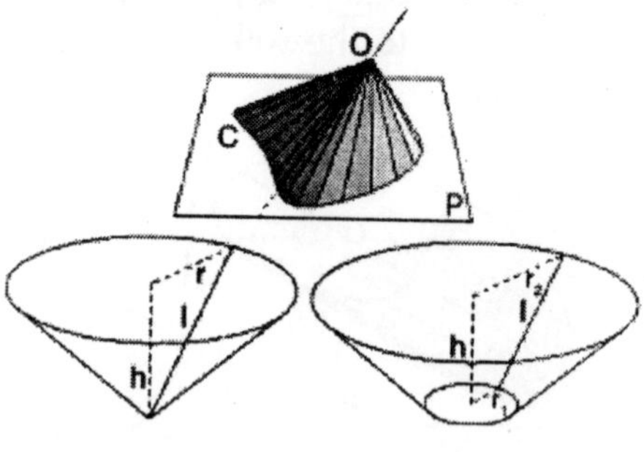

Fig. *Top:* A Cone with Vertex O and Directrix C. Bottom left: a right Circular Cone. Bottom right: *A* Frustum of the Latter.

The volume contained between P and the vertex O is,

$$V = \frac{1}{3} Ah,$$

where A is the area in the plane P enclosed by C and h is the from O and P (measured perpendicularly). The solid contained between P and a plane P' parallel to P (on the same side of the vertex) is called a frustum.

Its volume is,

$$V = \frac{1}{3}h(A + A' + \sqrt{AA'}),$$

where A and A' are the areas enclosed by the sections of the cone by *P* and *P′* (often called the bases of the frustum). The most important particular case of a cone is the right circular cone (often simply called a cone).

If *r* is the radius of the base, h is the altitude, and *l* is the length between the vertex and a point on the base circle, we have the following relations:

- *l*: $\sqrt{r^2 + h^2}$
- *Lateral area*: prl = pr $\sqrt{r^2 + h^2}$
- *Total area*: pr(*l* + *r*) = pr(*r* + $\sqrt{r^2 + h^2}$)
- *Volume*: $\frac{1}{3}\pi r^2 h$
- The implicit equation of this surface can be written
$$x^2 + y^2 = z^2;$$

For a frustum of right circular cone we have:

- $l\ \sqrt{r^2 + h^2}$, where $r = r^1 - y^2$
- *Lateral area*: $\pi(r^1 + r^2)l\sqrt{r^2 + h^2}$
- *Total area*: $\pi(r^{1^2} + r^{2^2} + (r^1 + r^2)l)$
- *Volume*: $\frac{1}{3}\pi r^2 h\ \ (r^{1^2} + r^{2^2} + r^1 r^2)$

SPHERES

The set of points in space whose distance to a fixed point (the centre) is a fixed positive number (the radius) is a sphere. A circle of radius r and centre (x^0, y^0, z^0) has equation,

$$(x - x^0) + (y - y)^2 + (z - z^0)^2 = r^2,$$

or,

$$x^2 + y^2 + z^2 - 2xx^0 - 2yy^0 - 2zz^0 + x^{02} + y^{02} + z^{02} - r^2 = 0.$$

Conversely, an equation of the form:

$$x^2 + y^2 + z^2 + 2dx + 2ey + 2fz + g = 0$$

defines a sphere if $d^2 + e^2 + f^2 > g$; the centre is $(-d, -e, -f)$ and the radius is,

$$\sqrt{d^2 + e^2 + f^2 - g}\,.$$

Four points not on the same plane determine a unique sphere. If the points have coordinates (x^1, y^1, z^1), (x^2, y^2, z^2), (x^3, y^3, z^3) and (x^4, x^4, z^4), the equation of the sphere is,

$$\begin{vmatrix} x^2 + y^2 + z^2 & x & y & z & 1 \\ x_2^1 + y_1^2 + z_2^2 & x_1 & y_1 & z_1 & 1 \\ x_2^2 + y_2^2 + z_2^2 & x_2 & y_2 & z_2 & 1 \\ x_3^3 + y_3^2 + z_3^2 & x_3 & y_3 & z_3 & 1 \\ x_4^2 + y_4^2 + z_4^2 & x_4 & y_4 & z_4 & 1 \end{vmatrix} = 0.$$

Given two points $P^1 = (x^1, y^1, z^1)$ and $P^2 = (x^2, y^2, z^2)$, there is a unique sphere whose diameter is P^1P^2; its equation is,

$$(x - x^1)(x - x^2) + (y - y^1)(y - y^2) + (z - z^1)(z - z^2) = 0.$$

The area of a sphere of radius r is $4\pi r^2$, and the volume is $\frac{4}{3}\pi r^3$. The area of a spherical polygon (that is, of a polygon on the sphere whose sides are arcs of great circles) is,

$$S = \left(\sum_{i=1}^{n} \theta_i - (n-2)\pi\right) r^2,$$

where r is the radius of the sphere, n is the number of vertices, and the θ_i are the internal angles of the polygons in radians. In particular, the sum of the angles of a spherical triangle is always greater than $\pi = 180°$, and the excess is proportional to the area.

CAP

Let the radius be r. The area of the curved region is $2\pi rh = \pi p^2$. The volume of the cap is,

$$\frac{1}{3}\pi h^2 (3r - h) = \frac{1}{8}\pi h(3a^2 + h^2).$$

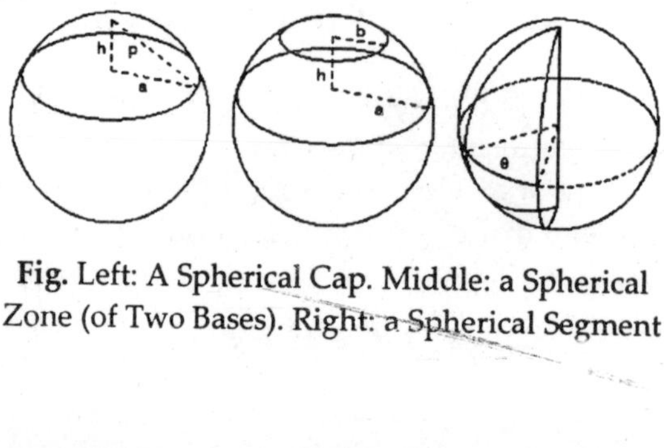

Fig. Left: A Spherical Cap. Middle: a Spherical Zone (of Two Bases). Right: a Spherical Segment

Spherical Zone

Let the radius be r. The area of the curved region is $2\pi rh$. The volume of the zone is,

$$\frac{1}{8}\pi h(3a^2 + 3b^2 + h^2).$$

Spherical Segment and Lune

Let the radius be r. The area of the curved region (lune) is $2r^2\theta$, the angle being measured in radians. The volume of the segment is,

$$\frac{2}{3}r^3\theta\,.$$

THE TORUS

A surface of revolution is formed by the rotation of a planar curve C about an axis in the plane of the curve and not cutting the curve.

The Pappus--Guldinus theorem says that:

- The area of the surface of revolution on a curve C is equal to the product of the length of C and the length of the path traced by the centroid of C (which is 2pthe distance from this centroid to the axis of revolution).
- The volume bounded by the surface of revolution on a simple closed curve C is equal to the product of the area bounded by C and the length of the path traced by the centroid of the area bounded by C.

When C is a circle, the surface obtained is a circular torus or torus of revolution. Let r be the radius of the revolving circle and let R be the distance from its centre to the axis of rotation. The area of the torus is $4\pi Rr^2$, and its volume is $2\pi Rr^2$.

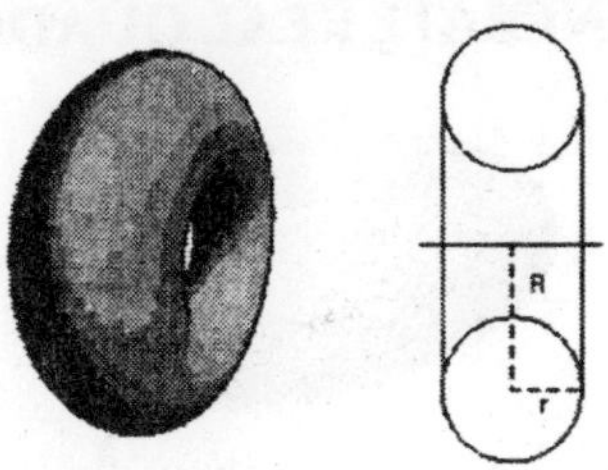

Fig. A Torus of Revolution

QUADRICS

A surface defined by an algebraic equation of degree two is called a quadric. Spheres, circular cylinders, and circular cones are quadrics.

By means of a rigid motion, any quadric can be transformed into a quadric having one of the following equations (where $a,b,c \neq 0$):

S. No.	Surfare Algebric Equation
1.	Real ellipsoid $x^2/a^2 + y^2/b^2 + z^2/c^2 = 1$
2.	Imaginary ellipsoid $x^2/a^2 + y^2/b^2 + z^2/c^2 = -1$
3.	Hyperboloid of one sheet $x^2/a^2 + y^2/b^2 - z^2/c^2 = 1$
4.	Hyperboloid of two sheets $x^2/a^2 + y^2/b^2 - z^2/c^2 = -1$
5.	Real quadric cone $x^2/a^2 + y^2/b^2 - z^2/c^2 = 0$
6.	Imaginary quadric cone $x^2/a^2 + y^2/b^2 + z^2/c^2 = 0$
7.	Elliptic paraboloid $x^2/a^2 + y^2/b^2 + 2z = 0$
8.	Hyperbolic paraboloid $x^2/a^2 - y^2/b^2 + 2z = 0$
9.	Real elliptic cylinder $x^2/a^2 + y^2/b^2 = 1$
10.	Imaginary elliptic cylinder $x^2/a^2 + y^2/b^2 = -1$
11.	Hyperbolic cylinder $x^2/a^2 - y^2/b^2 = 1$
12.	Real intersecting planes $x^2/a^2 - y^2/b^2 = 0$
13.	Imaginary intersecting planes $x^2 + 2y = 0$
14.	Parabolic cylinder $x^2 + 2y = 0$
15.	Real parallel planes $x^2 = 1$
16.	Imaginary parallel planes $x^2 = -1$
17.	Coincident planes $x^2 = 0$

FIVE NONDEGENERATE REAL QUADRICS

Fig. The Ellipsoid

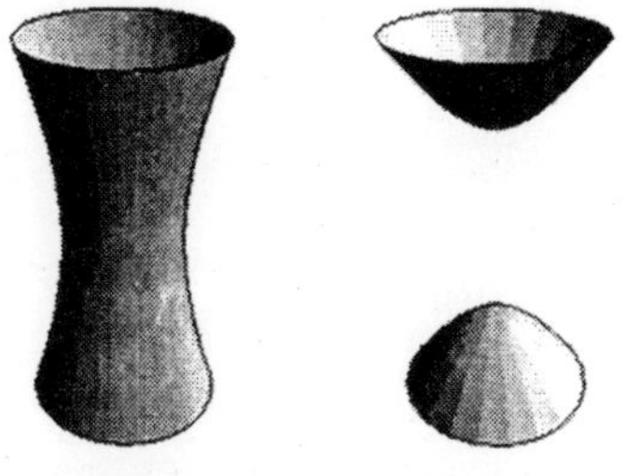

Fig. Left: Hyperboloid of One Sheet (3).
Right: Hyperboloid of Two Sheets (4).

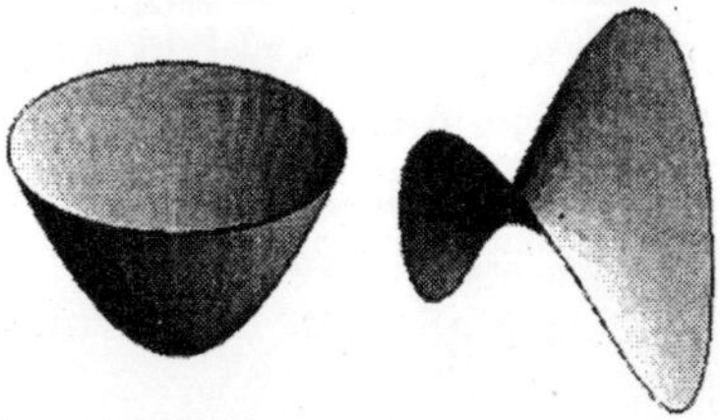

Fig. Left: Elliptic Paraboloid (7). Right: Hyperbolic Paraboloid (8).

Surfaces with equations (9) –(17) are cylinders over the planes curves of the same equation. Equations (2), (6), (10), (16), have no real solutions, so they do not describe surfaces in real three-dimensional space.

A surface with equation (5) can be regarded as a cone over a conic *C* (any ellipse, parabola or hyperbola can be taken as the directrix; there is a two-parameter family of essentially distinct cones over it, determined by the position of the vertex with respect to *C*).

The surfaces with equations (1) –(6) are central quadrics; in the form given, the centre is at the origin. The quantities a, b, c are the semiaxes. The volume of the ellipsoid with semiaxes *a, b, c* is,

$$\frac{4}{3}\pi abc .$$

When two of the semiaxes are the same, we can also write the area of the ellipsoid in closed form. Suppose $b = c$, so the ellipsoid $x^2/a^2 + (y^2 + z^2)/b^2 = 1$ is the surface of revolution obtained by rotating the ellipse $x^2/a^2 + y^2/b^2 = 1$ around the x-axis.

Its area is,

$$2\pi b^2 + \frac{2\pi a^2 b}{\sqrt{a^2 - b^2}} \arcsin \frac{\sqrt{a^2 - b^2}}{a} 2\pi b^2 + \frac{\pi a^2 b}{\sqrt{b^2 - a^2}} \log \frac{b + \sqrt{b^2 - a^2}}{b - \sqrt{b^2 - a^2}}.$$

The two quantities are equal, but only one avoids complex numbers, depending on whether a>b or a<b. When a>b, we have a prolate spheroid, that is, an ellipse rotated around its major axis; when a<b we have an oblate spheroid, which is an ellipse rotated around its minor axis.

Given a general quadratic equation in three variables,

$$ax^2 + by^2 + cz^2 + 2fyz + 2gzx + 2hxy + 2px + 2qy + 2rz + d = 0,$$

one can find out the type of conic it determines by consulting the following table:

ρ_2	ρ_4	Δ	k signs	K signs	Type of quadric
3	4	< 0			real ellipsoid
3	4	> 0	same		imaginary ellipsoid
3	4	> 0	opposite		hyperboloid of one sheet
3	4	< 0	opposite		hyperboloid of two sheets
3	3		opposite		real quadric cone
3	3		same		imaginary quadric cone
2	4	< 0	same		elliptic paraboloid
2	4	> 0	opposite		hyperbolic paraboloid
2	3		same	opposite	real elliptic cylinder
2	3		same	same	imaginary elliptic cylinder
2	3		opposite		hyperbolic cylinder
2	2		opposite		real intersecting planes
2	2		same		imaginary intersecting planes
1	3				parabolic cylinder
1	2		opposite		real parallel planes
1	2		same		1imaginary parallel planes
1	1				coincident planes

The columns have the following meaning.

Let,

$$e = \begin{bmatrix} a & h & g \\ h & b & f \\ g & f & c \end{bmatrix} \text{ and } E = \begin{bmatrix} a & h & g & p \\ h & b & f & q \\ g & f & c & f \\ p & q & r & d \end{bmatrix};$$

let ρ^3 and r^4 be the ranks of e and E, and let Δ be the determinant of E. The column "*k* signs" refers to the nonzero eigenvalues of e, that is, the roots of,

$$\begin{vmatrix} a-x & h & g \\ h & b-x & f \\ g & f & c-x \end{vmatrix} = 0'$$

if all nonzero eigenvalues have the same sign, choose "same", otherwise "opposite". Similarly, "*K* signs" refers to the sign of the nonzero eigenvalues of *E*.

CHARACTERISTICS COMPLETELY.

REPRESENTATION

In fact, a good representation should address the following issues:

- *Domain:* While no representation can describe all possible solids, a representation should be able to represent a useful set of geometric objects.
- *Unambiguity:* When you see a representation of a solid, you will know what is being represented without any doubt. An unambiguous representation is usually referred to as a complete one.
- *Uniqueness:* That is, there is only one way to represent a particular solid. If a representation is unique, then it is easy to determine if two solids are identical since one can just compare their representations.
- *Accuracy:* A representation is said accurate if no approximation is required.
- *Validness:* This means a representation should not

create any invalid or impossible solids. More precisely, a representation will not represent an object that does not correspond to a solid.

- *Closure:* Solids will be transformed and used with other operations such as union and intersection. "Closure" means that transforming a valid solid always yields a valid solid.
- *Compactness and Efficiency:* A good representation should be compact enough for saving space and allow for efficient algorithms to determine desired physical characteristics.

BOUNDARY REPRESENTATIONS

Boundary Representation, or B-rep for short, can be considered as an extension to the wireframe model. The merit of a B-rep is that a solid is bounded by its surface and has its interior and exterior. The surface of a solid consists of a set of well-organized faces, each of which is a piece of some surface (.e.g., a surface patch). Faces may share vertices and edges that are curve segments. Therefore, a B-rep is an extension to the wireframe model by adding face information to the latter.

There are two types of information in a B-rep: topological and geometric. Topological information provide the relationships among vertices, edges and faces similar to that used in a wireframe model.

In addition to connectivity, topological information also include orientation of edges and faces. Geometric information are usually equations of the edges and faces. The orientation of each face is important. Normally, a face is surrounded by a set of vertices. Using the right-handed rule, the ordering of these vertices for describing a particular face must guarantee that the normal vector of that face is pointing to the exterior of the solid. Normally, the order is counter clockwise.

If that face is given by an equation, the equation must be rewritten so that the normal vector at every point on the part that is being used as a face points to the exterior of the solid. Therefore, by inspecting normal vectors one can immediately tell the inside and outside of a solid under B-rep. This orientation must be done for all faces. The following shows three faces and their outward pointing normal vectors. To describe the top surface, the vertices should be 6, 7, 2, 1 or 7, 2, 1, 6 or 2, 1, 6, 7 or 1, 6, 7, 2. To describe the left face, the order should be 1, 2, 3, 4 or 2, 3, 4, 1 or 3, 4, 1, 2 or 4, 1, 2, 3.

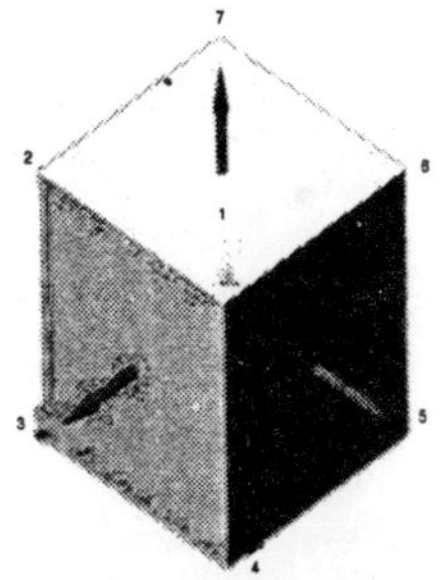

Unfortunately, not all surfaces can be oriented this way. If the surface of a solid can be oriented this way, it is called *orientable*; otherwise, it is non-*orientable*. The following shows the well-known *Mobius* band which is one-sided and non-*orientable*.

MANIFOLD CONDITIONS

The surface of a solid must satisfy some conditions so that the resulting solid is well-behaved. This is the so-called manifold condition. While recent research has shown that the surface of a solid does not have to be a manifold, we shall restrict to manifold surfaces to simply our discussion.

A surface is a 2-manifold if and only if for each point x on the surface there exists an open ball with centre x and sufficiently small radius so that the intersection of this ball and the surface can be continuously deformed to an open disk.

An open ball with centre at the coordinate origin and radius r is defined by equation $x^2 + y^2 + z^2 < r^2$. It contains all points inside of the sphere $x^2 + y^2 + z^2 = r^2$. An open disk is defined similarly $x^2 + y^2 < r^2$. By continuously deformed, it means one can twist or bend the shape without cutting (i.e. adjacency relations must be maintained) and gluing (i.e. a one-to-one relation is required). There is a cube and three open balls. Ball 2 has its centre on the top face. The intersection of this ball and the surface of the cube is an open disk (shown in red). Ball 1 has its centre on an edge. Its intersection with the surface of the cube is a "bent" open disk, which of course can be "unbent" to make it an open disk. Ball 3 has its centre at a corner. Its intersection with the cube's surface is a three-way bent open disk.

Let us take a look at the following figure:

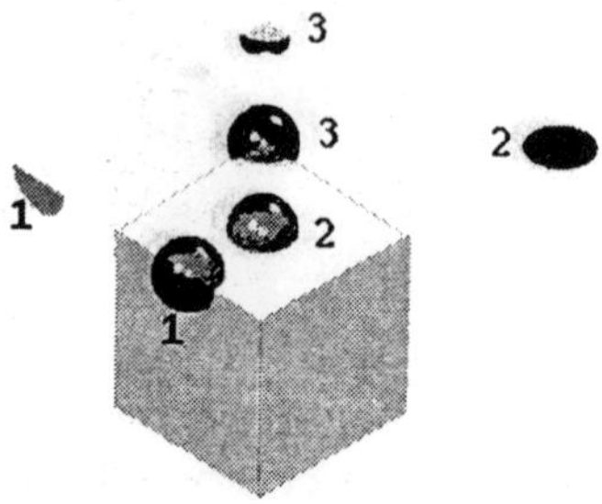

The following shows a solid whose bounding surface is not a manifold. The intersection of the open ball and the surface of the solid is the union of two intersecting open disks. This intersection cannot be deformed to an open disk without "gluing." Consequently, the surface is not a manifold.

WINGED-EDGE DATA STRUCTURE

Perhaps the oldest data structure for a B-rep is Baumgart's winged-edge data structure. It is quite different from that of a wireframe model, because the winged-edge data structure uses edges to keep track almost everything. *In what follows, we shall assume there is no holes in each face* and later extend it to cope with holes. Moreover, we shall assume edges and faces are line segments and polygons.

Topologically, one can always stretch curvilinear edges and faces so that they become flat without changing the relationships among them.

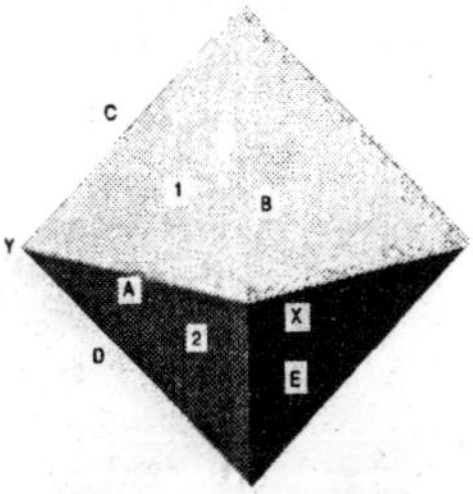

The above figure shows a polyhedron with vertices, edges and faces indicated with upper cases, lower cases and digits, respectively. Let us take a look at edge $a = XY$. This edge has two incident vertices X and Y, and two incident faces 1 and 2. A face is a polygon surrounded by edges. For example, face 1 has its edges *a, c* and *b*, and face 2 has its edges *a, e* and *d*. Note that the ordering is clockwise viewed from outside of the solid. If the direction of the edge is from X to Y, faces 1 and 2 are on the right and left side of edge *a*, respectively.

To capture the ordering of edges correctly, we need four more pieces of information. Since edge a is traversed once when traversing face 1 and traversed a second time when traversing face 2, it is used twice in different directions. For example, when traversing the edges of face 1, the predecessor and successor of edge a are edge *b* and edge *c*, and when traversing the edges of face 2, the predecessor and successor of edge a are edge *d* and edge *e*.

Note that although there are four edges incident to vertex X, only three of them are used when finding faces incident to edge *a*.

Therefore, for each edge, the following information are important:

- Vertices of this edge,
- Its left and right faces,
- The predecessor and successor of this edge when traversing its left face, and
- The predecessor and successor of this edge when traversing its right face.

FACES WITH HOLES

If some faces of a solid have holes, the above form of winged-edge data structure does not work. These holes may penetrate the solid (the left box below) or just like a pothole (the right box below).

There are two ways for resolving this problem:

1. For a face with inner loops, the outer boundary is ordered clockwise, while its inner loops, if any, are ordered counter clockwise.

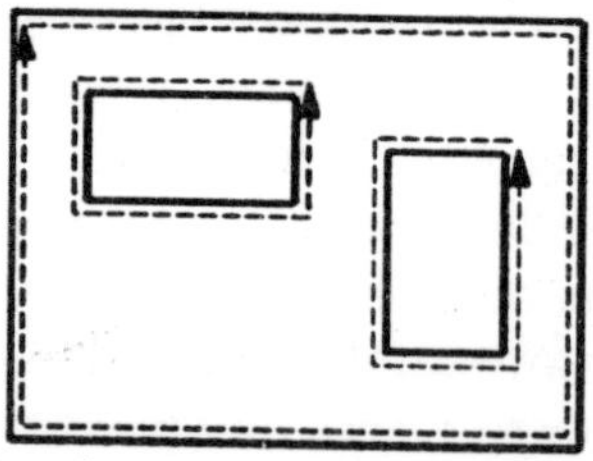

2. Another simple method is adding an auxiliary edge between each inner loop and the outer loop as shown below. This auxiliary edge will have the same face for its left and right faces. In this way, a face with holes becomes a single loop which can be represented with the winged-edge data structure. When traversing a loop, auxiliary edges can be identified easily since its left and right faces are the same.

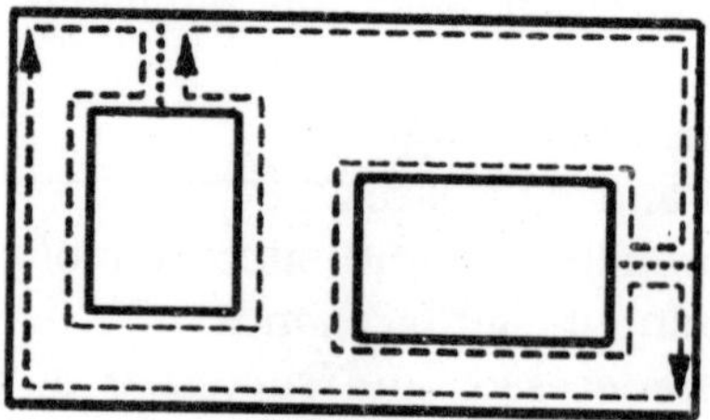

..........Auxiliary edges

EULER-POINCARE FORMULA

The Euler-Poincare formula describes the relationship of the number of vertices, the number of edges and the number of faces of a manifold. It has been generalized to include potholes and holes that penetrate the solid.

To state the Euler-Poincaré formula, we need the following definitions:

1. V: the number of vertices
2. E: the number of edges
3. F: the number of faces
4. G: the number of holes that penetrate the solid, usually referred to as genus in topology
5. S: the number of shells. A shell is an internal void of a solid. A shell is bounded by a 2-manifold surface, which can have its own genus value. Note that the solid itself is counted as a shell. Therefore, the value for S is at least 1.
6. L: the number of loops, all outer and inner loops of faces are counted.

Then, the Euler-Poincaré formula is the following:

$$V - E + F - (L - F) - 2(S - G) = 0$$

Examples:

1. A cube has eight vertices ($V = 8$), 12 edges ($E = 12$) and six faces ($F = 6$), no holes and one shell ($S = 1$); but $L = F$ since each face has only one outer loop. Therefore, we have,

$$V - E + F - (L - F) - 2(S - G) = 8 - 12 + 6 - (6 - 6) - 2(1 - 0) = 0$$

2. The following solid has 16 vertices, 24 edges, 11 faces, no holes, 1 shell and 12 loops (11 faces + one inner loop on the top face). Therefore,

$$V - E + F - (L - F) - 2(S - G) = 16 - 24 + 11 - (12 - 11) - 2(1 - 0) = 0$$

3. The following solid has 16 vertices, 24 edges, 10 faces, 1 hole (i.e., genus is 1), 1 shell and 12 loops (10 faces + 2 inner loops on top and bottom faces). Therefore,

$$V - E + F - (L - F) - 2(S - G) = 16 - 24$$
$$+ 10 - (12 - 10) - 2(1 - 1) = 0$$

4. The following solid has a penetrating hole and an internal cubic chamber as shown by the right cut-away figure. It has 24 vertices, 12*3 (cubes) = 36 edges, 6*3 (cubes)- 2 (top and bottom openings) = 16 faces, 1 hole (i.e., genus is 1), 2 shells and 18 loops (16 faces + 2 inner loops on top and bottom faces). Therefore,

$$V - E + F - (L - F) - 2(S - G) = 24 - 36$$
$$+ 16 - (18 - 16) - 2(2 - 1) = 0$$

5. The following solid has two penetrating holes and no internal chamber as shown by the right cut-away figure. It has 24 vertices, 36 edges, 14 faces, 2 hole (i.e., genus is 2), 1 shells and 18 loops (14 faces + 4). Therefore,

$$V - E + F - (L - F) - 2(S - G) = 24 - 36$$
$$+ 14 - (18 - 14) - 2(1 - 2) = 0$$

Part of the information recorded in a B-rep is topological (i.e., adjacency relations). Invalid solids may be generated if the representation is not carefully constructed. One way of checking this topological invalidity is to use the Euler-Poincaré formula. If its value is not zero, we are sure something must be wrong in the representation. However, this is only a one-side test. More precisely, a zero value of the Euler-Poincaré formula does not mean the solid is valid.

The figure above has a box and an additional sheet which is simply a rectangle. This object has 10 vertices, 15 edges, 7 faces, 1 shell and no hole. Its loop number is equal to the number of faces.

The value of the Euler-Poincaré formula is zero as shown below,

$$V - E + F - (L - F) - 2(S - G)$$
$$= 10 - 15 + 7 - (7 - 7) - 2(1 - 0) = 0$$

but this is not a valid solid! Therefore, if the value of Euler-Poincaré formula is non-zero, the representation is definitely not a valid solid. However, the value of the Euler-Poincaré formula being zero does not guarantee the representation would yield a valid solid.

CORRECT COUNTING OF GENUS

It is not always easy to count genus G correctly. For example, suppose we have a sphere punched by three tunnels as shown below. The left object shows the outside look. The right one shows the inside by cutting half of the sphere off. What is the genus value G? We know that the genus value counts the number of penetrating holes. But, in this example, it is somewhat ambiguous. In fact, by combining any two centre-going tunnels we will have three penetrating holes. However, this is incorrect!

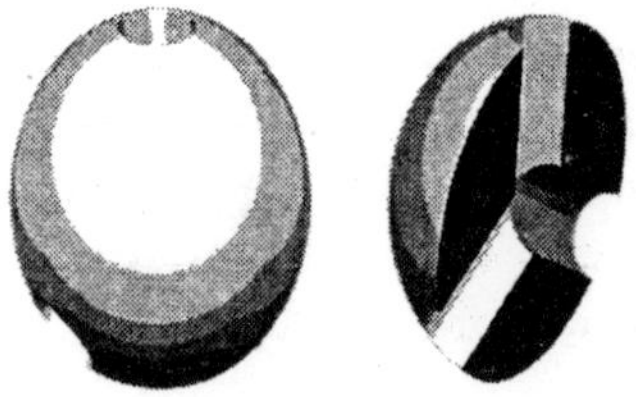

The Euler-Poincare Formula describes the topological property amount vertices, edges, faces, loops, shells and genus. Any topological transformation applied to the model will not alter this relationship.

Intuitively, applying topological transformations means we can twist, stretch and squash the model but we cannot cut some parts off nor glue some parts together. Let us apply some intuitive topological transformations to this model for computing the genus. We can push the "walls" surrounding the three tunnels so that the interior of the model becomes a thin shell. This is shown in the image below.

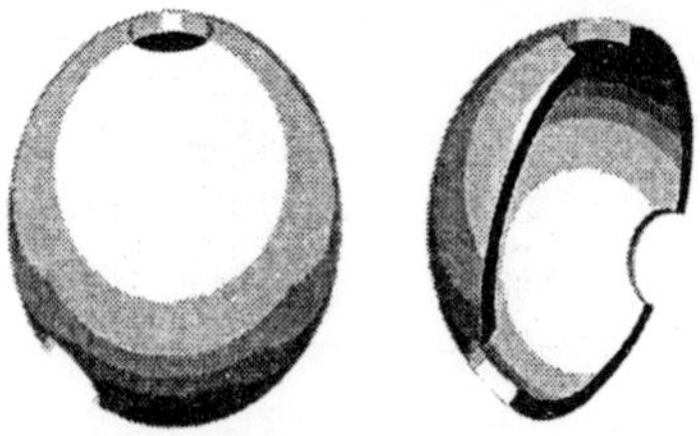

Stretch the top hole so that it is large enough (below left). Then, collapse the top portion to flatten the model. This is shown in the right model below. How many penetrating holes are there? Two! Therefore, G is 2.

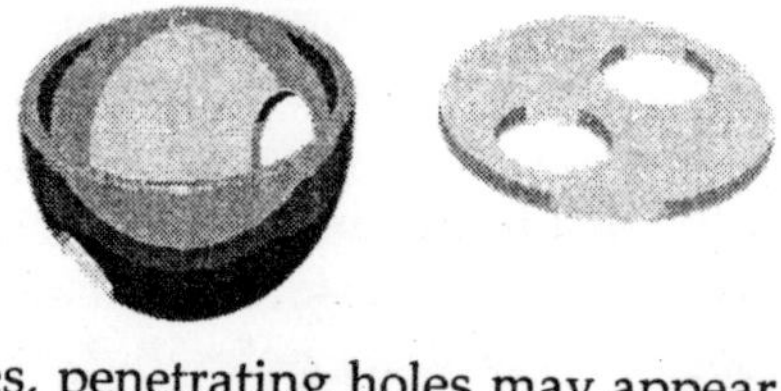

Sometimes, penetrating holes may appear in an unlikely situation. Consider the following model which is obtained by taking out a torus and tube from the interior of a sphere. What is the genus of this model?

It does not look like there is a penetrating hole. So, is genus equal to? In fact, $G = 1$! Figure it out yourself. Do some twisting, stretching and squashing.

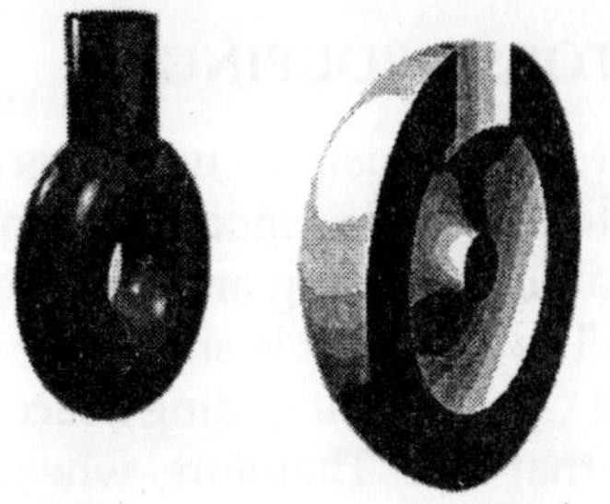

OPERATORS

Once a polyhedron model is available one might want to edit it by adding or deleting vertices, edges and faces to create a new polyhedron. These operations are called Euler Operators. However, it has been shown that in the process of editing a polyhedron with Euler operators, some intermediate results may not be valid solids at all.

Recall from the discussion of the Euler-Poincaré formula that the following holds for all polyhedra:

$$V - E + F - (L - F) - 2(S - G) = 0$$

where V, E, F, L, S and G are the numbers of vertices, edges, faces, loops, shells and genus, respectively. Based on this relation, some Euler operators have been selected for editing a polyhedron so that the Euler-Poincaré formula is always satisfied.

There are two groups of such operators: the Make group and the Kill group. Operators start with M and K are operators of the Make and Kill groups, respectively. Euler operators are written as Mxyz and Kxyz for operations in the Make and Kill groups, respectively, where x, y and z are elements of the model (e.g., a vertex, edge, face, loop, shell and genus). For example, MEV means adding an edge and a vertex while KEV means deleting an edge and a vertex.

It has been proved by Mantyla in 1984 that Euler operators form a complete set of modeling primitives for manifold solids. More precisely, every topologically valid polyhedron can be constructed from an initial polyhedron by a finite sequence of Euler operators. Therefore, Euler operators are powerful operations.

EULER OPERATORS GROUPING

The Make group consists of four operators for adding some elements into the existing model creating a new one, and a Make-Kill operator for adding and deleting some elements at the same time. The above table shows the change of values of *V, E, F, L, S* and G. Note that adding a face produces a loop, the outer loop of that face. Therefore, when F is increased, L should also be increased. This new loop and the new face will cancel each other in the subexpression *L* – *F*. Please verify that none of these operators would cause the Euler-Poincaré formula to fail. The following table illustrates the way of using Euler operators to construct a tetrahedron. Vertices, edges, and faces are in red, blue and green; the only shell is in transparent gray. The first step uses *MSFV* to obtain a shell with a face and a vertex. The next three steps use *MEVs,* each of which adds an edge and a vertex. The last three steps use *MFE,* each of which adds a face and an edge. Thus, in seven steps or seven Euler operators a tetrahedron is constructed.

Operator Name	Meaning	V	E	F	L	S	G	Result
MSFV	Make a shell, a face and a vertex	+1		+1	+1	+1		
MEV	Make an edge and a vertex	+1	+1					
MEV	Make an edge and a vertex	+1		+1				
MEV	Make an edge and a vertex	+1	+1					

MFE	Make a face						
	and an edge	+1		+1	+1		
MFE	Make a face						
	and an edge	+1	+1		+1		
MFE	Make a face						
	and an edge	+1	+1		+1		

MSG simply makes a shell with a hole. After this, one can add vertices, edges, faces, loops. There must be loops, because the new hole penetrates at least one face.

MEKL makes an edge and at the same time kills a loop. A commonly used *MEKL* is adding an edge connecting the outer loop and the inner loop of a face. In this case, the number of edges E is increased by 1 and the number of loops L is decreased by 1 since that loop is killed. The left figure below shows two loops of the top face while the right one shows the new edge added after performing a *MEKL*. Thus, the inner and outer are "joint" together with the new edge becoming a single face.

CONSTRUCTIVE SOLID GEOMETRY

Constructive Solid Geometry, or CSG for short, is yet another way of representing solids. A CSG solid is constructed from a few primitives with Boolean operators (i.e., set union, intersection and difference). Thus, a CSG solid can be written as a set equations and can also be considered a design methodology.

CSG PRIMITIVES

The standard CSG primitives consist of the block (i.e.,

cube), triangular prism, sphere, cylinder, cone and torus. These six primitives are in some normal or generic form and must be instantiated by the user to be used in his/her design. Moreover, the instantiated primitive may require transformations such as scaling, translation and rotation to be positioned at the desired place.

Suppose the block primitive is defined by its "lower left" corner < –1, –1, –1 > and "upper right" corner < 1, 1, 1 >. To produce a rectangular box with centre at < 3, 2, 3 > and height and width 3 and length 5, a user may first scale the block primitive 1.5 times in the y- and z-direction and 2.5 times in the x-direction, and then translate the result to < 3, 2, 3 >. If the block primitive is called Block in a CSG system, the result may be obtained as follows: translate(scale(Block, < 2.5, 1.5, 1.5 >), < 3, 2, 3 >). In the above, the object to be transformed and the transformation data are the first and second arguments, respectively.

BOOLEAN OPERATORS

We can combined two instantiated and perhaps transformed primitives into one with set union, set intersection and set difference operators. However, simple set operators may create problems as will be discussed in regularized Boolean operators, modifications are required. Let us just use set operations on this page.

Given two sets, *A* and *B*, its union consists of all points from either *A* or *B*; its intersection consists of all points in both sets; and its difference, written as *A* – *B* (resp., *B* – *A*), consists of all points in *A* but not in *B* (resp., in *B* but not in *A*). In the following, A is the vertical cylinder and *B* is the horizontal cylinder. From left to right, the four solids are the union and intersection of *A* and *B*, *A* – *B* and *B* – *A*.

Therefore, a solid can be considered as the result of applying Boolean operators to a set of instantiated and transformed CSG primitives. Let us take a look at a simple example. We want to design a bracket-like shape with a hole shown on the right-most figure below. We start with two instantiations of blocks and one instantiation of a cylinder (the left-most figure). Then, the two blocks are scaled and one of them is rotated to a vertical position. The cylinder is also scaled so that its radius matches that of the hole. These three instantiations are than transformed to their desired positions. The final product is obtained by computing the union of the two blocks and then subtracting from it the cylinder.

CENTER OF MASS AND MOMENT OF INERTIA

MASS

We saw before that the double integral over a region of the constant function 1 measures the area of the region. If the region has uniform density 1, then the mass is the density times the area which equals the area. What if the density is not constant. Suppose that the density is given by the continuous function. Density = $\rho(x,y)$. In this case we can cut the region into tiny rectangles where the density is approximately constant.

The area of mass rectangle is given by,

$$\text{Mass} = (\text{Density})(\text{Area}) = \rho(x,y)\,\Delta x\,\Delta y$$

If we add all to masses together and take the limit as the rectangle size goes to zero, we get a double integral.

Mass: Let $\rho(x,y)$ be the density of a lamina (flat sheet) R at the point (x,y). Then the total mass of the lamina is the double integral

$$Mass = \iint_R \rho(x,y)\,dydx$$

Example: A rectangular metal sheet with $2 < x < 5$ and $0 < y < 3$ has density function $\rho(x,y) = x + y$. Set up the double integral that gives the mass of the metal sheet.

Solution: We just have the integral.

$$\int_2^5 \int_0^3 (x+y)\,dydx$$

MOMENTS AND CENTER OF MASS

We have seen that in calculus the moments about an axis are defined by the product of the mass times the distance from the axis.

- M_x = (Mass)(y)
- M_y = (Mass)(x)

If we have a region R with density function r(x,y), then we do the usual thing. We cut the region into small rectangles for which the density is constant and add up the moments of each of these rectangles. Then take the limit as the rectangle size approaches zero. This will give us the total moment.

Moments of Mass and Center of Mass: Suppose that ρ(x,y) is a continuous density function on a lamina R. Then the moments of mass are,

$$M_x = \iint_R \rho(x,y)\, y\,dydx \qquad M_y = \iint_R \rho(x,y)\, x\,dydx$$

and if M is the mass of the lamina, then the center of mass is,

$$(\bar{x},\bar{y}) = \left(\frac{M_y}{M}, \frac{M_x}{M}\right)$$

Example: Set up the integrals that give the center of mass of the rectangle with vertices (0,0), (1,0), (1,1), and (0,1) and density function proportional to the square of the distance from the origin. Use a calculator or computer to evaluate these integrals.

Solution: The mass is given by,

$$M = \int_0^1\int_0^1 k\left(x^2 + y^2\right) dydx = \frac{2k}{3}$$

The moments are given by,

$$M = \int_0^1\int_0^1 k\left(x^2 + y^2\right) ydydx \text{ and } M = \int_0^1\int_0^1 k\left(x^2 + y^2\right) xdydx$$

These evaluate to,

- M_x = 5k/12
- M_y = 5k/12

It should not be a surprise that the moments are equal since there is complete symmetry with respect to x and y. Finally, we divide to get (x,y) = (5/8,5/8). This tells us that the metal plate will balance perfectly if we place a pin at (5/8,5/8)

MOMENTS OF INERTIA

We often call M_x and M_y the first moments. They have first powers of y and x in their definitions and help find the center of mass. We define the *moments of inertia* (or second moments) by introducing squares of y and x in their definitions. The moments of inertia help us find the kinetic energy in rotational motion. Below is the definition

Moments of Inertia: Suppose that ρ(x,y) is a continuous density function on a lamina R. Then the *moments of inertia* are,

$$I_x = \iint_R \rho(x,y)\, y^2 dydx \quad I_y = \iint_R \rho(x,y)\, y^2 dydx$$

SURFACE AREA

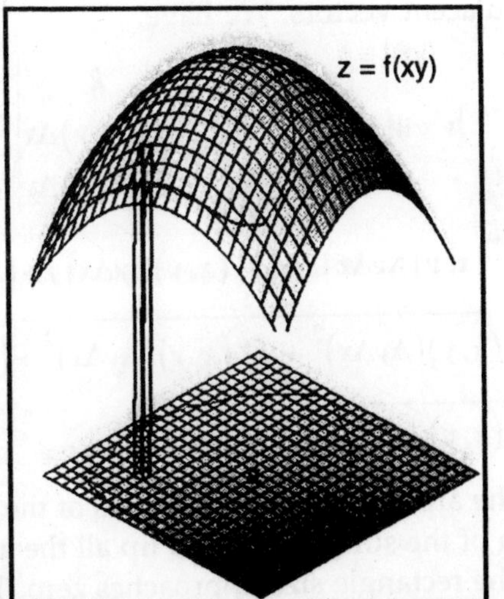

We can think of a smooth surface as a quilt flapping in the wind. It consists of many rectangles patched together. More generally and more accurately, let z = f(x,y) be a surface in R^3 defined over a region Rin the xy-plane. cut the xy-plane into rectangles. Each rectangle will project vertically to a piece of the surface as shown in the figure below. Although the area of the rectangle in R is,

$$\text{Area} = \Delta y \Delta x$$

The area of the corresponding piece of the surface will not be DyDx since it is not a rectangle. Even if we cut finely, we will still not produce a rectangle, but rather will approximately produce a parallelogram. With a little geometry we can see that the two adjacent sides of the parallelogram are (in vector form),

$$u = \Delta x\, i + f_x(x,y)\Delta x\, k \text{ and } v = f_y(x,y)\Delta y\, i + \Delta y\, k.$$

We can see this by realizing that the partial derivatives are the slopes in each direction. If we run Δx in the i direction, then we will rise $f_x(x,y)\Delta x$ in the k direction so that rise/run = $f_x(x,y)$. Which agrees with the slope idea of the partial derivative. A similar argument will confirm the equation for the vector v.

Now that we know the adjacent vectors we recall that the area of a parallelogram is the magnitude of the cross product of the two adjacent vectors. We have,

$$|v \times w| = \begin{vmatrix} i & j & k \\ \Delta x & 0 & f_x(x,y)\Delta x \\ 0 & \Delta_y & f_y(x,y)\Delta y \end{vmatrix}$$

$$= \left|-\left(f_y(x,y)\Delta y\Delta x\right)i - \left(f_x(x,y)\Delta y\Delta x\right)j + \left(\Delta y\Delta x\right)k\right|$$

$$= \sqrt{f_y^2(x,y)(\Delta y\Delta x)^2 + f_x^2(x,y)(\Delta y\Delta x)^2 + (\Delta y\Delta x)^2}$$

$$= \sqrt{f_y^2(x,y) + f_x^2(x,y) + 1\Delta y\Delta x}$$

This is the area of one of the patches of the quilt. To find the total area of the surface, we add up all the areas and take the limit as the rectangle size approaches zero. This results in a double Riemann sum, that is a double integral. We state the definition below.

Definition of Surface Area: Let z = f(x,y) be a differentiable surface defined over a region R.

Then its surface area is given by,

$$Surface\ Area = \iint_R = \sqrt{1+f_x^2(x,y)f_y^2(x,y)dydx}$$

Example: Find the surface area of the part of the plane, z = 8x + 4y that lies inside the cylinder $x^2 + y^2 = 16$.

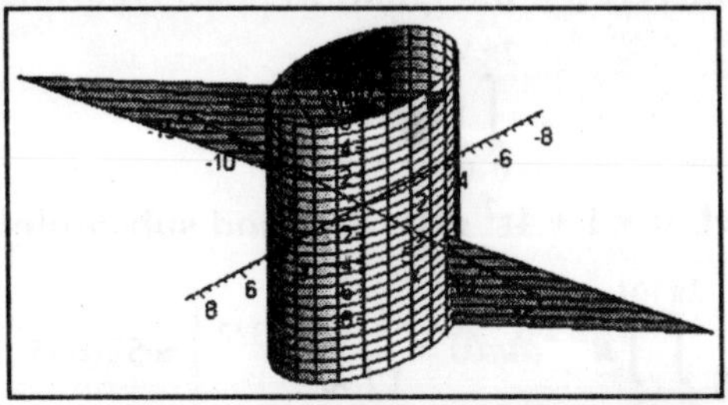

Solution: We calculate partial derivatives, $f_x(x,y) = 8$ $f_y(x,y) = 4$, so that $1 + f_x^2(x,y) + f_y^2(x,y) = 1 + 64 + 16 = 81$.

Taking a square root and integrating, we get,

$$\iint_R 9dydx$$

We could work this integral out, but there is a much easier way. The integral of a constant is just the constant times the area of the region. Since the region is a circle, we get,

Surface Area = 9(16π) = 144π

In reality, since there is a square root in the formula, most surface area calculations require intensive integration skills or the use of a machine. The prior example and the next example are not meant to deceive, but rather to show how the essence of surface area problems work without the integration difficulty clouding your understanding.

Example: Find the surface area of the part of the paraboloid, $z = 25 - x^2 - y^2$ that lies above the xy-plane.

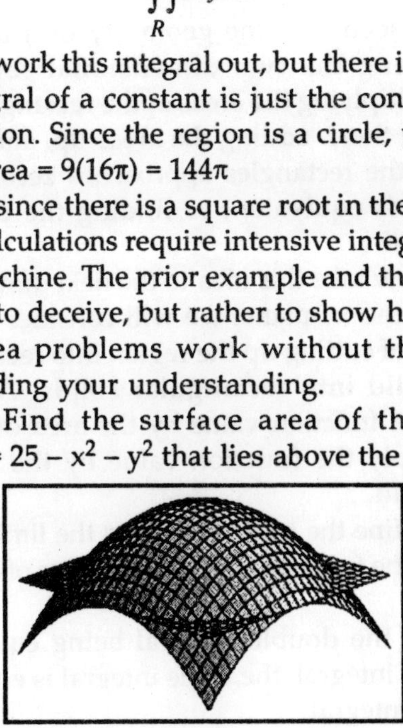

Solution: We calculate partial derivatives, $f_x(x,y) = -2x$ $f_y(x,y) = -2y$, so that $1 + f_x^2(x,y) + f_y^2(x,y) = 1 + 4x^2 + 4y^2$.

At this point if we listen closely, we should hear a little voice pleading "Polar Coordinates". We listen to its call and realise that the region is just the circle r = 5.

Now convert the integrand to polar coordinates to get,

$$\int_0^{2\pi}\int_0^5 \sqrt{1+4r^2}\, r\,dr\,d\theta$$

Now let, $u = 1 + 4r^2$ $du = 8rdr$ and substitute,

$$\frac{1}{8}\int_0^{2\pi}\int_1^{101} u^{1/2}\,du\,d\theta = \frac{1}{12}\int_0^{2\pi} u^{3/2}\Big|_1^{101} \approx 530.95$$

TRIPLE INTEGRALS

DEFINITION OF THE TRIPLE INTEGRAL

We have seen that the geometry of a double integral involves cutting the two dimensional region into tiny rectangles, multiplying the areas of the rectangles by the value of the function there, adding the areas up, and taking a limit as the size of the rectangles approaches zero. We have also seen that this is equivalent to finding the double iterated iterated integral.

We will now take this idea to the next dimension. Instead of a region in the xy-plane, we will consider a solid in xyz-space. Instead of cutting up the region into rectangles, we will cut up the solid into rectangular solids. And instead of multiplying the function value by the area of the rectangle, we will multiply the function value by the volume of the rectangular solid.

We can define the *triple integral* as the limit of the sum of the product of the function times the volume of the rectangular solids.

Instead of the double integral being equivalent to the double iterated integral, the triple integral is equivalent to the triple iterated integral.

Definition of the Triple Integral: Let f(x,y,z) be a continuous function of three variables defined over a solid Q. Then the triple integral over Q is defined as,

$$\iiint_Q f(x,y,z) = \lim \sum f(x,y,z)\Delta x \Delta y \Delta z$$

where the sum is taken over the rectangular solids included in the solid Q and lim is taken to mean the limit as the side lengths of the rectangular solid.

This definition is only practical for estimating the triple integral when a data set is given. When we have a symbolically defined function, we use an extension of the fundamental theorem of calculus which is just Fubini's theorem for triple integrals.

Theorem for Evaluating Triple Integrals: Let f(x,y,z) be a continuous function over a solid Q defined by,

$a < x < b$
$h_1(x) < y < h_2(x)$
$g_1(x,y) < z < g_2(x,y)$

Then the triple integral is equal to the triple iterated integral.

$$\iiint_Q f(x,y,z) = \int_a^b \int_{k_1(x)}^{k_2(x)} \int_{g_1(x,y)}^{g_2(x,y)} f(x,y,z)\,dzdydx$$

Remark: As with double integrals the order of integration can be changed with care.

EXAMPLE

Evaluate:

$$\iiint_Q f(x,y,z)\,dzdydx$$

Where,

$$f(x,y,z) = 1 - x$$

and Q is the solid that lies in the first octant and below the plain,

$$3x + 2y + z = 6$$

Solution: The picture of the region,

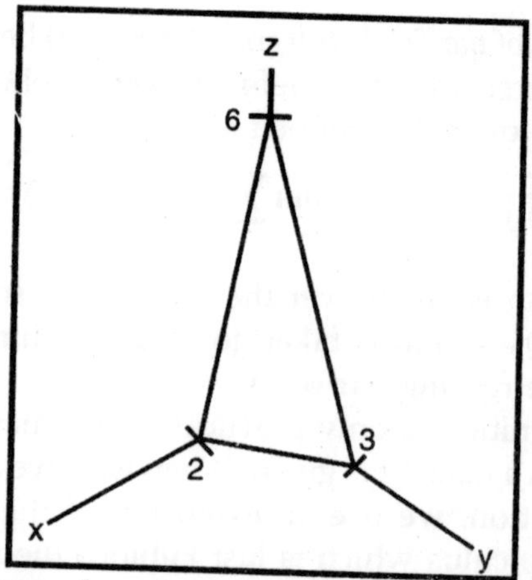

The challenge here is to find the limits. We work on the innermost limit first which corresponds with the variable "z". Think of standing vertically. Your feet will rest on the lower limit and your head will touch the higher limit. The lower limit is the xy-plane or $z = 0$. The upper limit is the given plane. Solving for z, we get, $z = 6 - 3x - 2y$. Now we work on the middle limits that correspond to the variable "y". We look at the projection of the surface in the xy-plane. It is shown below.

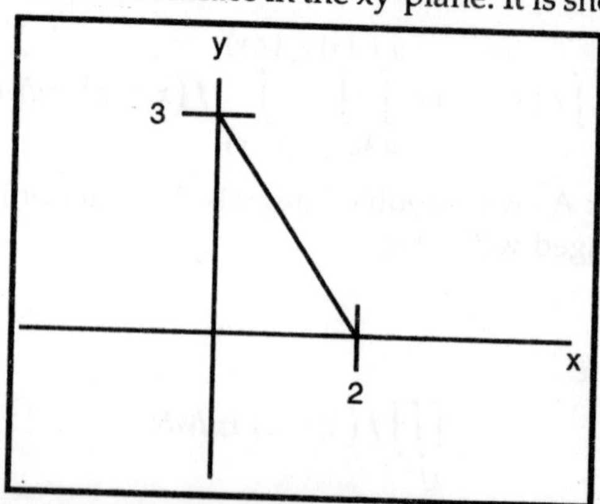

Now we find the limits just as we found the limits of double integrals. The lower limit is just, $y = 0$. If we set $z = 0$ and solve for y, we get for the upper limit, $y = 3 - 3/2\, x$.

Next we find the outer limits, corresponding to the variable "x". The lowest x gets is 0 and highest x gets is 2.

Hence, $0 < x < 2$. The integral is thus,

$$\int_0^2\int_0^{3-\frac{3}{2}x}\int_0^{6-3x-2y}(1-x)\,dzdydx=\int_0^2\int_0^{3-\frac{3}{2}x}\left[z-xz\right]_0^{6-3x-2y}dydx'$$

$$=\int_0^2\int_0^{3-\frac{3}{2}x}\left[(6-3x-2y)-(6x-3x^2 2xy)\right]dydx$$

$$=\int_0^2\left[\left(6y-9xy-y^2+3x^2y+xy^2\right)\right]_0^{3-\frac{3}{2}x}dx$$

$$=\int_0^2\left(18-9x-27x+\frac{27}{2}x^2-9x-\frac{9}{4}x^2+9x^2-\frac{9}{2}x^3+9x-9x^2+\frac{9}{4}x^3\right)dx$$

$$=\int_0^2\left(9-18x+\frac{45}{4}x^2-\frac{9}{4}x^3\right)dx$$

$$=\left[9x-9x^2+\frac{15}{4}x^3-\frac{9}{16}x^4\right]_0^2=3$$

Example: Switch the order of integration from the previous example so that dydxdz appears.

Solution: This time we work on the "y" variabie first. The lower limit for the y-variable is 0. For the upper limit, we solve for y in the plane to get, $y = 3 - 3/2\,x - 1/2\,z$

To find the "x" limits, we project onto the xz-plane as shown below,

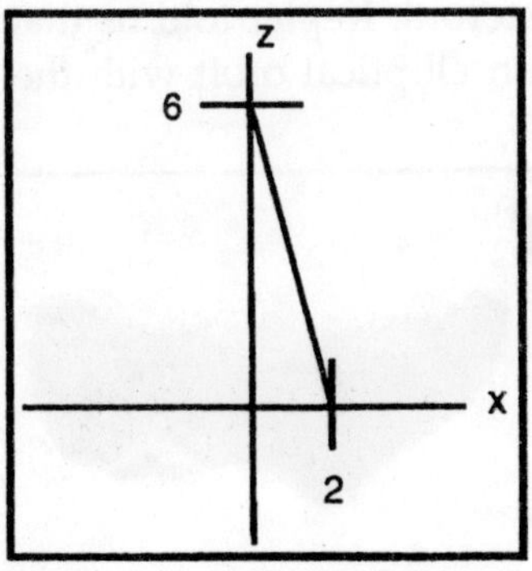

The lower limit for x is 0. To find the upper limit we set y

= 0 and solve for x to get, x = 2 – 1/3 z. Finally, to get the limits for z, we see that the smallest z will get is 0 and the largest z will get is 6. We get, 0 < z < 6.

We can write,

$$\int_0^6 \int_0^{2-\frac{1}{3}x} \int_0^{6-\frac{3}{2}x-\frac{1}{3}y} (1-x)\,dydxdz$$

MASS, CENTER OF MASS, AND MOMENTS OF INERTIA

For a three dimensional solid with constant density, the mass is the density times the volume. If the density is not constant but rather a continuous function of x, y, and x, then we can cut the solid into very small rectangular solids so that on each rectangular solid the density is approximately constant. The volume of the rectangle is,

ΔMass = (Density)(ΔVolume) = f(x,y,z) Δx Δy Δz

Now do the usual thing: We add up all the small masses and take the limit as the rectangular solids get small. This will give us the triple integral

$$Mass = \iiint_Q f(x,y,z)\,dzdydx$$

We are often interested in the center of mass of a solid. For example when the NEAR satellite orbited around the asteroid Eros, NASA scientists needed to compute the center of mass of the asteroid. Kepler told us that a stable orbit will always orbit in an elliptical orbit with the center of mass as one of the foci.

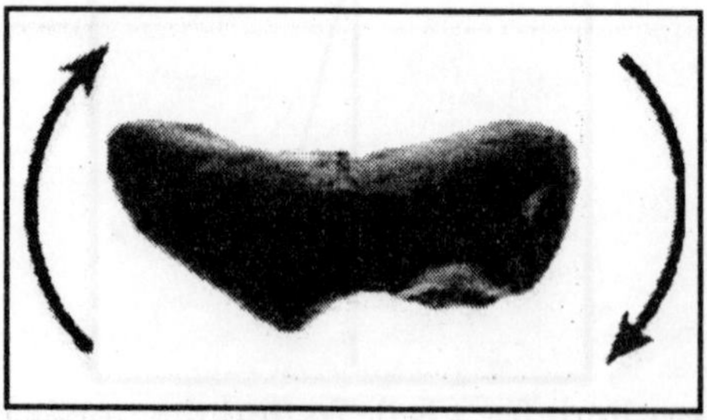

The NEAR satellite orbiting around Eros: We find the center of mass of a solid just as we found the center of mass of a lamina. Since we are in three dimensions, instead of the moments about the axes, we find the moments about the coordinate planes. We state the definitions from physics below.54

Moments and Center of Mass: Let ρ(x,y,z) be the density of a solid Q. Then the *first moments* about the coordinate planes are,

$$M_{yz} = \iiint_Q x\rho(x,y,z)\,dzdydx$$

$$M_{xz} = \iiint_Q y\rho(x,y,z)\,dzdydx$$

$$M_{xy} = \iiint_Q z\rho(x,y,z)\,dzdydx$$

and the *center of mass* is given by,

$$(\overline{x},\overline{y},\overline{z}) = \left(\frac{Myz}{M},\frac{Mxz}{M},\frac{Mxy}{M}\right)$$

Notice that letting the density function being identically equal to 1 gives the volume,

$$Volume = \iiint_Q dzdydx$$

Example: Find the center of mass of the solid that lies below the paraboloid $z = 4 - x^2 - y^2$ that lies above the xy-plane if the density of the region is given by $\rho(x,y,z) = x^2 + 2y^2 + z$.

You may use your calculator or computer to evaluate the integrals. Just as with lamina, there are formulas for moments of inertial about the three axes. They involve multiplying the density function by the square of the distance from the axes. We have

Moments of Inertia: Let ρ(x,y,z) be the density of a solid Q. Then the *first moments of inertia* about the coordinate axes are

$$I_x = \iiint_Q (y^2 + z^2)\rho(x,y,z)\,dzdydx$$

$$I_y = \iiint_Q (x^2 + z^2)\rho(x, y, z)\, dzdydx$$

$$I_z = \iiint_Q (x^2 + y^2)\rho(x, y, z)\, dzdydx$$

TRIPLE INTEGRALS IN CYLINDRICAL AND SPHERICAL COORDINATES

CYLINDRICAL COORDINATES

When we were working with double integrals, we saw that it was often easier to convert to polar coordinates. For triple integrals we have been introduced to three coordinate systems. The rectangular coordinate system (x,y,z) is the system that we are used to. The other two systems, cylindrical coordinates (ρ,θ,z)and spherical coordinates (ρ,θ,ϕ) are the topic of this discussion.

Recall that cylindrical coordinates are most appropriate when the expression, $x^2 + y^2$ occurs. The construction is just an extension of polar coordinates.

x = r cos θ y = r sin θ z = z

Since triple integration can be looked at as iterated integration we have,

$$\int_a^b \int_{h_1(x)}^{h_2(x)} \int_{g_1(x,y)}^{g_2(x,y)} f(x, y, z)\, dzdydx$$

$$= \int_a^b \int_{h_1(x)}^{h_2(x)} \left[\int_{g_1(x,y)}^{g_2(x,y)} f(x, y, z)\, dz \right] dydx$$

$$= \int_{\theta_1}^{\theta_2} \int_{r_1(\theta)}^{r_2(\theta)} \left[\int_{g_1(r\cos\theta, r\sin\theta)}^{g_2(r\cos\theta, r\sin\theta)} f((r\cos\theta, r\sin\theta z))\, dz \right] rdrd\theta$$

$$= \int_{\theta_1}^{\theta_2} \int_{r_1(x)}^{r_2(x)} \int_{g_1(r\cos\theta, r\sin\theta)}^{g_2(r\cos\theta, r\sin\theta)} f((r\cos\theta, r\sin\theta, z))\, rdzdrd\theta$$

This leads us the the following theorem

Theorem: Integration With Cylindrical Coordinates

Let f(x,y,z) be a continuous function on a solid Q. Then,

$$\iiint_Q f(x,y,z)\,dzdydx = \iiint_Q f(r\cos\theta, r\sin\theta, z)\,rdzdrd\theta$$

Example: Find the moment of inertia about the z-axis of the solid that lies below the paraboloid, $z = 25 - x^2 - y^2$ inside the cylinder, $x^2 + y^2 = 4$, above the xy-plane, and has density function, $r(x,y,z) = x^2 + y^2 + 6z$.

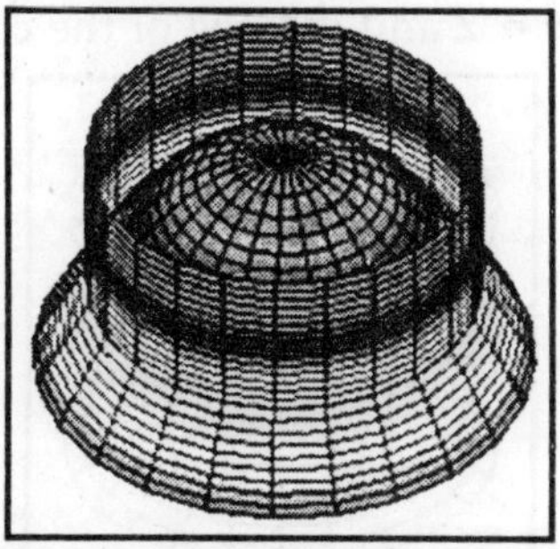

Solution: By the moment of inertia formula, we have,

$$I_x = \iiint_Q f\left(x^2+y^2\right)\left(x^2+y^2+2z\right)dzdydx =$$

The region, being inside of a cylinder is ripe for cylindrical coordinates. We get,

$$I_x = \int_0^{2\pi}\int_0^{2}\int_0^{25-r^2} r^2\left(r^2+6z\right)rdzdrd\theta = \int_0^{2\pi}\int_0^{2}\left[r^5z+3r^3z^2\right]_0^{25-r^2} drd\theta$$

$$\int_0^{2\pi}\int_0^{2}\left[-125r^5+2r^7+1875r^3\right]drd\theta \frac{37384\pi}{3}$$

SPHERICAL COORDINATES

Another coordinate system that often comes into use is the spherical coordinate system.

To review, the transformations are,

$x = \rho\cos\theta\sin\phi$ $\qquad$ $y = \rho\sin\theta\sin\phi$ $\qquad$ $z = \rho\cos\phi$

In the next section we will show that, dzdydx = $\rho^2 \sin\phi \, d\rho d\phi d\theta$.

Theorem: Integration With Spherical Coordinates. Let f(x,y,z) be a continuous function on a solid Q. Then

$$\iiint_Q f(x,y,z)\,dzdydx =$$

$$\iiint_Q f(\rho\cos\theta\sin\phi, \rho\sin\theta\sin\phi, \cos\phi)\rho^2 \sin\phi d\rho d\phi d\theta$$

Example: Find the volume of solid that lies inside the sphere, $x^2 + y^2 + z^2 = 2$ and outside of the cone, $z^2 = x^2 + y^2$.

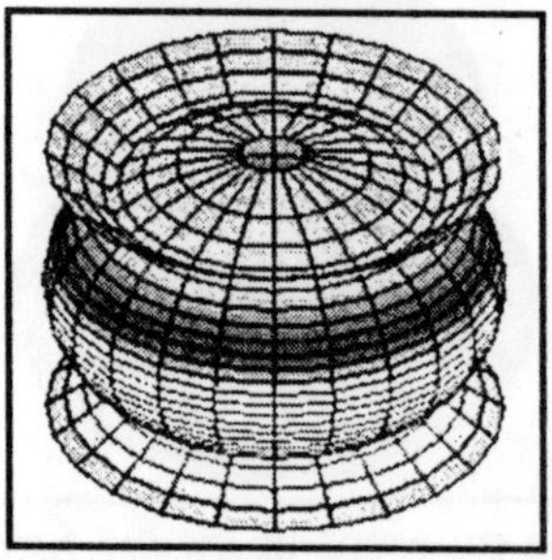

Solution: We convert to spherical coordinates. The sphere becomes $\rho = \sqrt{2}$. To convert the cone, we add z^2 to both sides of the equation $2z^2 = x^2 + y^2 + z^2$. Now convert to $2\rho^2\cos^2\phi = \rho^2$. Canceling the ρ^2 and solving for f we get, $\phi = \cos^{-1}(1/) = \pi/4$ or $7\pi/4$. In spherical coordinates (since the coordinates are p periodic) $7\pi/4 = 3\pi/4$.

To find the volume we compute:

$$V = \int_0^{2\pi} \int_{\frac{\pi}{4}}^{\frac{3\pi}{4}} \int_0^{\sqrt{2}} \rho^2 \sin\phi d\rho\, d\phi\, d\theta$$

Evaluating this integral should be routine at this point and is equal to:

$$V = \frac{8\pi}{3}$$

Jacobians: Review of the Idea of Substitution.

Consider the integral:

$$\int_0^2 x\cos\left(x^2\right)dx$$

To evaluate this integral we use the u-substitution, $u = x^2$. This substitution sends the interval [0,2] onto the interval [0,4]. We can see that there is stretching of the interval. The stretching is not uniform. In fact, the first part [0,0.5] is actually contracted. This is the reason why we need to find du.

$$\frac{du}{dx} = 2x \, or \, \frac{dx}{du} = \frac{1}{2x}$$

This is the factor that needs to be multiplied in when we perform the substitution. Notice for small positive values of x, this factor is greater than 1 and for large values of x, the factor is smaller than 1. This is how the stretching and contracting is accounted for.

JACOBIANS

We have seen that when we convert to polar coordinates, we use dydx = rdrdθ. With a geometrical argument, we showed why the "extra r" is included. Taking the analogy from the one variable case, the transformation to polar coordinates produces stretching and contracting. The "extra r" takes care of this stretching and contracting. The goal for this section is to be able to find the "extra factor" for a more general transformation. We call this "extra factor" the *Jacobian* of the transformation. We can find it by taking the determinant of the two by two matrix of partial derivatives.

Definition of the Jacobian: Let x = g(u,v) andy = h(u,v) be a transformation of the plane. Then the Jacobian of this transformation is,

$$\frac{\partial(x,y)}{\partial(u,v)} = \begin{vmatrix} \frac{\partial x}{\partial u} & \frac{\partial x}{\partial v} \\ \frac{\partial y}{\partial u} & \frac{\partial y}{\partial v} \end{vmatrix} - \frac{\partial x}{\partial u}\frac{\partial y}{\partial v} - \frac{\partial x}{\partial v}\frac{\partial y}{\partial u}$$

Example: Find the Jacobian of the polar coordinates transformation x(r,θ) = r cos θ y(r,θ) = r sin θ.

Solution: We have,

$$\frac{\partial(x,y)}{\partial(r,\theta)} = \begin{vmatrix} \cos\theta & -r\sin\theta \\ \sin\theta & -r\cos\theta \end{vmatrix} = r\cos^2\theta + r\sin^2\theta = r$$

This is comforting since it agrees with the extra factor in integration.

DOUBLE INTEGRATION AND THE JACOBIAN

Theorem: Integration and Coordinate Transformations.

Let, $T: R^2 \to R^2$ given by x = g(u,v), y = h(u,v) be a transformation on the plane that is one to one from a region S to a region R. If g and h have continuous partial derivatives such that the Jacobian is never zero, then,

$$\iint_R f(x,y)\,dydx = \iint_S f(g(u,v),h(u,v))\,dydx = \left|\frac{\partial(x,y)}{\partial(u,v)}\right| dudv$$

Remark: A useful fact is that the Jacobian of the inverse transformation is the reciprocal of the Jacobian of the original transformation.

$$\left|\frac{\partial(x,y)}{\partial(u,v)}\right| = \frac{1}{\left|\frac{\partial(u,u)}{\partial(x,y)}\right|}$$

This is a consequence of the fact that the determinant of the inverse of a matrix A is the reciprocal of the determinant of A.

IDEA OF THE PROOF

As usual, we cut S up into tiny rectangles so that the image under T of each rectangle is a parallelogram.

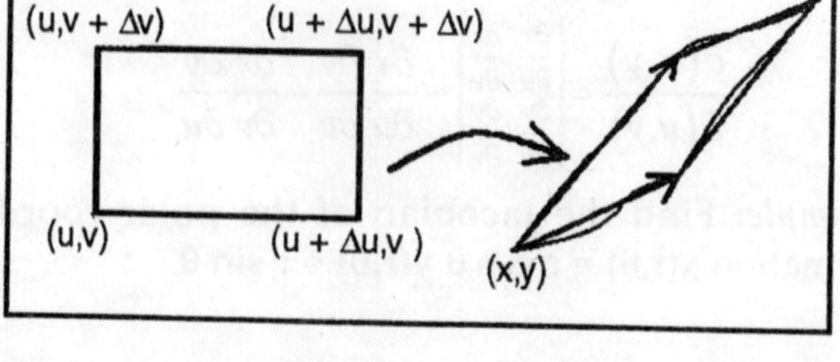

We need to find the area of the parallelogram.

Considering differentials, we have:

$T(u + \Delta u, v) \cong T(u,v) + (x_u \Delta u, y_u \Delta u)$

$T(u, v + \Delta v) \cong T(u,v) + (x_v \Delta v, y_v \Delta v)$

Thus the two vectors that make the parallelogram are:

$P = g_u \Delta u\, i + h_u \Delta u\, j$

$Q = g_v \Delta v\, i + h_v \Delta v\, j$

To find the area of this parallelogram we just cross the two vectors.

$$P \times Q = \begin{vmatrix} i & j & k \\ x_u \Delta u & y_u \Delta u & 0 \\ x_u \Delta u & y_u \Delta u & 0 \end{vmatrix} = \left| x_u y_u - x_u y_u \right| \Delta u \Delta v = \left| \frac{\partial(x,y)}{\partial(u,v)} \right| \Delta u \Delta v$$

and the extra factor is revealed.

Example: Use an appropriate change of variables to find the volume of the region below $z = (x - y)^2$ above the x-axis, over the parallelogram with vertices (0,0), (1,1), (2,0), and (1,– 1).

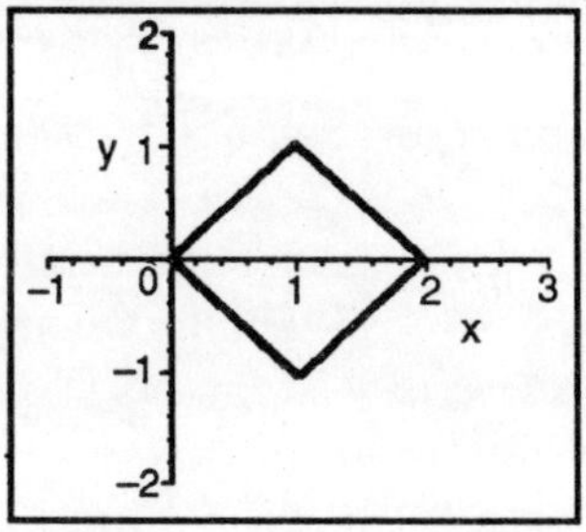

Solution: We find the equations of the four lines that make the parallelogram to be

$y = x \qquad y = x - 2 \qquad y = -x \qquad y = -x + 2$

or

$x - y = 0 \qquad x - y = 2 \qquad x + y = 0 \qquad x + y = 2$

The region is given by: $0 < x - y < 2$ and $0 < x + y < 2$. This leads us to the inverse transformation,

$$u(x,y) = x - y \quad v(x,y) = x + y$$

The Jacobian of the inverse transformation is,

$$\left| \frac{\partial(u,v)}{\partial(x,y)} \right| = \begin{vmatrix} 1 & -1 \\ 1 & 1 \end{vmatrix} = 2$$

Since the Jacobian is the reciprocal of the inverse Jacobian we get,

$$\left|\frac{\partial(x,y)}{\partial(u,v)}\right| = \frac{1}{2}$$

The region is given by, 0 < u < 2 and 0 < v < 2 and the function is given by z = u^2.

Putting this all together, we get the double integral:

$$\int_0^2\int_0^2 u^2\left(\frac{1}{2}\right)dudv = \int_0^2\left[\frac{u^3}{6}\right]_0^2 dv$$

$$\int_0^2 \frac{4}{3}dv = \frac{8}{3}$$

10

Limits

RATIO TESTS

There are also very useful tests involving the ratio v_{n+1}/v_n of two successive terms of a series. In these tests, we must assume that u_n and v_n are *strictly* positive.

Assume that $u_n > 0$, $v_n > 0$ and,

$$\frac{vn+1}{v_n} \leqq \frac{un+1}{u_n} \tag{1}$$

for sufficiently large n, say for $n \geq n_0$.

Then,

$$v_n = \frac{v_{n_0+1}}{v_{n_0}}\frac{v_{n_0+2}}{v_{n_0+1}}\cdots\frac{v_n}{v_{n-1}}v_{n_0} \leqq \frac{u_{n_0+1}}{u_{n_0}}\frac{u_{n_0+2}}{u_{n_0+1}}\cdots\frac{u_n}{u_{n-1}}v_{n_0} = \frac{v_{n_0}}{u_{n_0}}u_n,$$

so that v_n £ Ku_n, where K is independent of n. Similarly,

$$\frac{v_{n+1}}{v_n} \geqq \frac{u_{n+1}}{u_n} \tag{2}$$

for $n \geq n_0$ implies $v_n \geq Ku_n$, with a positive K. Hence

1. If (1) is true for sufficiently large n and Su_n converges, then Sv_n converges.
2. If (2) is true for sufficiently large n and Σu_n diverges, then Σv_n diverges. Setting in Theorem 4 $u_n = r^n$, we find
3. The series v_n converges if $v_{n+1}/v_n \leq r$, where $r < 1$ for all sufficiently large n.

   This test is known as d'Alembert's test. The corresponding divergence test that Σv_n diverges, if $v_{n+1}/v_n \geq r$, where $r \geq 1$ for all sufficiently large n, is trivial.

We shall see later that d'Alembert's test is theoretically less general than Cauchy's in that Cauchy's test can always be applied when d'Alembert's test can be applied, and often when it cannot; cf., for example. Ratio tests are useless for 'irregular' series such as

$$0+\frac{1}{2}+0+\frac{1}{4}+0+\frac{1}{8}+\ldots.$$

Nonetheless, d'Alembert's test is very useful in practice, because, when v_n is a complicated function, v_{n+1}/v_n is often much less complicated and so easier to work with.

It often happens that v_{n+1}/v_n or $v^{1/n}{}_n$ tends to a limit as $n \to \infty$*. When this limit is less than 1, it is evident that the conditions of Theorems 2 or 6 above are satisfied.

4. *If* $v^{1/n}{}_n$ or v_{n+1}/v_n tends to a limit less than unity when $n \to \infty$, then Σv_n converges.

It is almost obvious that if either function tends to a limit larger than unity, then $S\, v_n$ diverges. We leave the formal proof of this as an exercise to the reader. But when $v^{1/n}{}_n$ or v_{n+1}/v_n tends to 1 these tests fail. They fail also when $v^{1/n}{}_n$ or v_{n+1}/v_n oscillates in such a way that, while always less than 1, it assumes for an infinity of values of n values approaching indefinitely near to 1; the tests which involve v_{n+1}/v_n fail also when that ratio oscillates so as to be sometimes less than and sometimes larger than 1. When $v^{1/n}{}_n$ behaves in this way, Theorem 3 is sufficient to prove the divergence of the series. But it is clear that there is a wide margin of cases in which more subtle tests will be needed.

D' ALEMBERT'S RATIO

We have explained that what was meant by saying that an infinite series is *converges, diverges* or is *oscillatory* and illustrated our definitions by a few simple examples, mainly derived from the geometrical series,

$$1+x+x^2+\ldots$$

and other series connected with it. In this chapter, we shall pursue the subject in a more systematic manner and prove several theorems which enable us to determine when the simplest series, which occur commonly in analysis, are convergent. We shall often use the notation,

$$u_m + u_{m+1} + \cdots + u_n = \sum_{m}^{n} u_y,$$

and write,

$$\sum_{0}^{\infty} u_n, or\ simply\ \Sigma u_n,$$

for the infinite series,

$$u_0 + u_1 + u_2 + ...^*.$$

It is immaterial whether we denote our series by $u_1 + u_2 + ...$ or by $u_0 + u_1 + ...$ (as here). Later we shall be concerned with series of the type $a_0 + a_1x + a_2x^2 + ...$, for which the latter notation is clearly more convenient. We shall.therefore adopt this as our standard notation. However, we shall not adhere to it systematically and shall assume that u_1 is the first term whenever this course is more convenient. For example, it is more convenient when dealing with the series $1 + 1/2 + 1/3 + ...$ to assume that $u_n = 1/n$ and the series begins with u_1 than to assume that $u_n = 1/(n + 1)$ and the series begins with u_0.

SERIES OF POSITIVE TERMS

The theory of the convergence of series is comparatively simple when all the terms of the series considered are positive. We shall consider such series first not only because they are the easiest to deal with, but also because the discussion of the convergence of a series containing negative or complex terms can often be made to depend upon a similar discussion of a series of positive terms only.

Here and in what follows, 'positive' is to be regarded as including zero. When we are discussing the convergence or divergence of a series, we may disregard any finite number of terms. Thus, when a series contains a finite number only of negative or complex terms, we may omit them and apply the theorems which follow to the remainder.

COMPARISON THEOREM

If Σ_{un} converges and $v_n \le u_n$ for all values of n, then Σ_{vn} converges and $\Sigma_{vn} \le \Sigma_{un}$. Moregenerally, if $v_n \le Ku_n$, where K is a constant, then Σ_{vn} converges and $\Sigma_{vn} \le K\Sigma_{un}$. And if Σ_{un} diverges and vn $\le Ku_n$ with apositive K, then Σ_{vn} diverges.

Moreover, in inferring the convergence or divergence of $\Sigma\, v_n$ by means of one of these tests, it is sufficient to know that the test is satisfied for *sufficiently large* values of *n*, i.e., for all values of *n* larger than a definite value n_0. However, of course, the conclusion that $\Sigma\, v_n \ge KS\, u_n$ does not necessarily hold in this case.

A particularly useful case of this theorem is *If* $S\, u_n$ converges (diverges) and u_n/v_n tends to a limit other than zero as $n \to \infty$, then Σv_n converges (diverges).

FIRST APPLICATIONS OF THESE TESTS

The most important theorem we have proved about the convergence of any particular series is that $S\, r_n$ converges if $r <$ 1 and diverges if r ³ 1*, and it is natural to take $u_n = r^n$ in Theorem C.

We find:

1. The series Σv_n converges if $v_n \le Kr^n$, where $r < 1$, for all sufficiently large values of n.

   Throughout this chapter, *r* is positive in the wider sense which includes zero.

   When $K = 1$, this condition may be written in the form $v_n^{1/n} \le r$, whence we obtain what is known as *Cauchy's test* for the convergence of a series of positive terms, viz.,
2. The series Sv_n converges if $v_n^{1/n}$ £ r, where $r < 1$, for all sufficiently large values of n. On the other hand,
3. The series Σv_n diverges if $v_n^{1/n} \ge 1$ for an infinity of values n.

   This is obvious because $v_n^{1/n} \ge 1$ involves $v_n \ge 1$.

DIRICHLET'S THEOREM

Before proceeding further in the investigation of tests of

convergence and divergence, we shall prove an important general theorem concerning series of positive terms.

DIRICHLET'S THEOREM

The sum of a series of positive terms is the same in whatever order the terms are taken. This theorem seems to have been stated explicitly first by Dirichlet in 1837. It was no doubt known to earlier writers, and in particular to Cauchy. This theorem asserts that, if we have a convergent series of positive terms, say,

$$u_0 + u_1 + u_2 + \ldots$$

and form any other series

$$v_0 + v_1 + v_2 + \ldots$$

out of the same terms by taking them in any new order, then the second series converges and has the same sum as the first. Of course, no terms must be omitted: Every u must come somewhere among the v's and *vice versa*.

The proof is extremely simple. Let s be the sum of the series of u's. Then the sum of any number of terms selected from the u's is not larger than s. But every v is a u, whence the sum of any number of terms selected from the u's is not larger than s.

Hence, Σv_n converges and its sum t is not larger than s. But we can show in exactly the same way that $s \leq t$.

Thus,

$$s = t.$$

MULTIPLICATION OF SERIES OF POSITIVE TERMS

An immediate corollary of Dirichlet's theorem is the theorem: If,

$$u_0 + u_1 + u_2 + \ldots \text{ and } v_0 + v_1 + v_2 + \ldots$$

are two convergent series of positive terms and s, t are their respective sums, then the series,

$$u_0v_0 + (u_1v_0 + u_0v_1) + (u_2v_0 + u_1v_1 + u_0v_2) + \ldots$$

converges and has the sum st. Arrange all the possible products of pairs $u_m v_n$ in the form of a doubly infinite array,

u_0v_0	u_1v_0	u_2v_0	u_3v_0	...
u_1v_0	u_1v_1	u_2v_1	u_3v_1	...
u_0v_2	u_1v_2	u_2v_2	u_3v_2	...
u_0v_3	u_1v_3	u_2v_3	u_3v_3	...
...	...	...	...	...

We can rearrange these terms in the form of a simply infinite series in a variety of ways.

Among these are the following:

- We begin with the single term u_0v_0 for which $m + n = 0$; then, we take the two terms u_1v_0, u_0v_1 for which $m + n = 1$; then,the three terms u_2v_0, u_1v_1, u_0v_2 for which $m + n = 2$; etc. We thus obtain the series: $u_0v_0 + (u_1v_0 + u_0v_1) +(u_2v_0 +u_1 v_1 + u_0v_2) +\ldots$ of the theorem.
- We begin with the single term u_0v_0 for which both suffixes are zero; then we take the terms u_1v_0, u_1v_1, u_1v_0, u_0v_1 which involve a subscript 1, but no higher suffix; then the terms u_0v_0, u_2v_1, u_2v_2, u_1v_2, u_0v_2 which involve a suffix 2 but no higher suffix; etc.

The sums of these groups of terms, respectively, are equal to and the sum of the first $n + 1$ groups is,

$$u_0v_0 \; (u_0 + u_1) +(v_0 +v_1) -u_0v_0,$$

$$(u_0 + u_1 + u_2) +(v_0 + v_1 + v_2) - (u_0 + u_1) +(v_0 + v_1)$$

and tends to *st* as $n \to \infty$. When the sum of the series is formed in this manner, the sum. of the first one, two, three... groups comprises all the terms in the first, second, third... rectangles indicated.

The sum of the series formed in the second manner is *st*. But the first series is (when the brackets are removed) a rearrangement of the second, whence, by Dirichlet's theorem, it converges to the sum *st*. Thus, the theorem is proved.

ABEL'S (OR PRINGSHEIM'S) THEOREM

If Σu_n is a convergent series of positive and decreasing terms, then lim $nu_n = 0$.

In fact, $u_{n+1} + u_{n+2} + \ldots \to 0$ and, a fortiori, $u_{n+1} + u_{n+2} + \ldots + u_{2n} \to 0$, and the left-hand side is at least nu_{2n}, whence $2nu_{2n} = 2(nu_{2n}) \to 0$.

Also,

$$(2n+1)u_{2n+1} \leqq \frac{2n+1}{2n} 2nu_{2n} \to 0,$$

whence $u_n \to 0$.

Examples:

1. Use Abel's theorem to show that Σn^{-1} and $\Sigma(an+b)^{-1}$ diverge.
2. Show that Abel's theorem is not true if we omit the condition that u_n decreases as n increases.
   [The series
   $$1+\frac{1}{2^2}+\frac{1}{3^2}+\frac{1}{4}+\frac{1}{5^2}+\frac{1}{6^2}+\frac{1}{7^2}+\frac{1}{8^2}+\frac{1}{9}+\frac{1}{10^2}+\ldots,$$
   in which $u_n = 1/n$ or $1/n^2$, according to whether n is or is not a perfect square, converges, since it may be rearranged in the form,
   $$\frac{1}{2^2}+\frac{1}{3^2}+\frac{1}{5^2}+\frac{1}{6^2}+\frac{1}{7^2}+\frac{1}{8^2}+\frac{1}{10^2}+\ldots+\left(1+\frac{1}{4}+\frac{1}{9}+\ldots\right),$$
   and each of these series converges. However, since $nu_n = 1$ whenever n is a perfect square, it is not true that $nu_n \to 0$.]
3. The converse of Abel's theorem is not true, i.e., it is not true that, if u_n decreases with n and $\lim nu_n = 0$, then Σu_n converges.
   [Take the series Σn^{-1} and multiply the first term by 1, the second by 1/2, the next two terms by 1/3, the next four terms by 1/4, the next eight by 1/5, etc. On grouping the terms of the new series thus formed in brackets, we obtain
   $$1+\frac{1}{2}\cdot\frac{1}{2}+\frac{1}{3}\left(\frac{1}{3}+\frac{1}{4}\right)+\ldots\frac{1}{4}\left(\frac{1}{5}+\frac{1}{6}+\frac{1}{7}+\frac{1}{8}\right)\ldots;$$
   and this series diverges, since its terms are larger than those of,

$$1+\frac{1}{2}\cdot\frac{1}{2}+\frac{1}{3}\cdot\frac{1}{2}+\frac{1}{4}\cdot\frac{1}{2}...,$$

which diverges. But it is easy to see that the terms of the series

$$1+\frac{1}{2}\cdot\frac{1}{2}+\frac{1}{3}\cdot\frac{1}{3}+\frac{1}{3}\cdot\frac{1}{4}+\frac{1}{4}\cdot\frac{1}{5}+\frac{1}{4}\cdot\frac{1}{6}+...$$

satisfy the condition that $nu_n \to 0$. In fact, $nu_n = 1/n$ if $2^{n-2} < n$ £ 2^{n-1} and $n \to \infty$ as $n \to \infty$.]

MACLAURIN'S (OR CAUCHY'S) INTEGRAL TEST

The test was discovered by Maclaurin and redsacovered by Cauchy, to whom it is often attributed. If u_n decreases steadily as n increases, we can write $u_n = \phi(n)$ and assume that $\phi(n)$ is the value assumed, when $x = n$, by a continuous and steadily decreasing function $\phi(x)$ of the continuous variable x. Then, if n is any positive integer, we have,

$$\phi(v-1) \geqq \phi(x) \geqq \phi(v)$$

when $n-1 \leq x \leq n$. Let,

$$u_v = \phi(v-1) - \int_{v-1}^{v} \phi(x)\,dx = \int_{v-1}^{v} \{\phi(v-1) - \phi(x)\}dx,$$

so that,

$$0 \leqq u_v \leqq \phi(v-1) - \phi(v).$$

Then, Σv_n is a series of positive terms and,

$$v_2 + v_2 + ... + v_n \leqq \phi(1) - \phi(n) \leqq \phi(1).$$

Hence Σv_n converges and so,

$$u_2 + u_2 + ... + v_n \; or \sum_{1}^{n-1} \phi(v) - \int_1^n \phi(x)dx$$

tends to a positive limit, not exceeding $\phi(1)$, when $n \to \infty$.

Let us write,

$$\Phi(\xi) = \int_1^{\xi} \phi(x)dx,$$

so that $\Phi(\xi)$ is a continuous and steadily increasing function of ξ.. Then,

$$u_1 + u_2 + ... + u_{n-1} - \Phi(n)$$

tends to a positive limit, not exceeding $\phi(1)$ when $n \to \infty$. Hence, Σu_n converges or diverges according to whether $F(n)$ tends to a limit or to infinity as $n \to \infty$, and therefore, since $\Phi(n)$ increases steadily, according to whether $\Phi(\xi)$ tends to a limit or to infinity as $\xi \to \infty$. Hence, if $\phi(x)$ is a function of x which is positive and continuous for all values of x larger than unity and decreasessteadily as x increases, then he series,

$$\phi(1) + \phi(2) + \ldots$$

does or does not converge according to whether,

$$\Phi(\xi) = \int_1^\xi \phi(x)\,dx$$

does or does not tend to a limit l as $x \to \infty$; in the first case, the sum of the series is not larger than $\phi(1) + l$.

In fact, the sum must be less than $\phi(1) + l$, that $v_n < \phi(n-1) - \phi(n)$ unless $\phi(x) = \phi(n)$ throughout the interval $(\nu-1, \nu)$ and this cannot be true for all values of ν.

THE SERIES ΣN⁻

By far the most important application of the integral test is to the series,

$$1^{-8} + 2^{-8} + 3^{-8} + \ldots + n^{-8} + \ldots,$$

where s is any rational number. We have seen already that the series diverges when $s = 1$. If $s \le 0$, it is obvious that the series diverges. If $s > 0$, then u_n decreases as n increases and we can apply the test. Here,

$$\Phi(\xi) = \int_1^\xi \frac{dx}{x^8} = \frac{\xi 1^{-8} - 1}{1 - 8},$$

unless $s = 1$. If $s > 1$, then $\xi^{1-S} \to 0$ as $\xi \to \infty$, and thus,

$$\Phi(\xi) \to \frac{1}{8-1} = 1,$$

say. And if $s < 1$, then $\xi^{1-S} \to \infty$ as $\xi \to \infty$, and so $\Phi(\xi) \to \infty$. Thus, *the series Σn^{-S} converges if $s > 1$, diverges if $s \ge 1$ and in the first case its sum is less than $s/(s-1)$.*

Naturally, we could prove that the series diverges when $s < 1$ by comparing it with the divergent series Σn^{-1}.

However, it is interesting to see how the integral test may be applied to the series Σn^{-1} when the preceding analysis fails.

In this case,

$$\Phi(\xi) = \int_1^{\xi} \frac{dx}{x},$$

and it is easy to see that $\Phi(\xi) \to \infty$ when $\xi \to \infty$. In fact, if $\xi > 2^n$, then,

$$\Phi(\xi) > \int_1^{2^n} \frac{dx}{x} = \int_1^{2} \frac{dx}{x} + \int_2^{4} \frac{dx}{x} + \ldots + \int_{2^{n-1}}^{2^n} \frac{dx}{x}$$

However, by setting $x = 2^r u$, we obtain,

$$\int_{2^r}^{2^{n+1}} \frac{dx}{x} = \int_1^{2} \frac{du}{u}, \textit{and so}\, \Phi(\xi) > n \int_2^{2} \frac{du}{u},$$

which shows that $\Phi(\xi) \to \infty$ when $\xi \to \infty$.

Examples:

1. Prove by an argument similar to that used above and without integration, that

$$\Phi(\xi) = \int_1^{\xi} \frac{dx}{x8},$$

where $s < 1$, tends to infinity with ξ.

2. The series

$$\Sigma n^{-2}, \Sigma n - \frac{3}{2}, \Sigma n - \frac{11}{10}$$

converge and their sums are not larger than 2, 3, 11, respectively. The series

$$\Sigma n - \frac{1}{2}, \Sigma n - \frac{11}{10}$$

diverge.

3. The series

$$\Sigma \frac{n^s}{n^t + a},$$

where a > 0, converges or diverges according to whether $t > 1 + s$ or $t \leq 1 + s$. [Compare with Σn^{s-t}.]

4. Discuss the convergence or divergence of the series,

$$\Sigma \frac{a_1 n^{s_1} + a_2 n^{s_2} + \ldots + a_k n^{s_k}}{b_1 n^{t_1} + b_2 n^{t_2} + \ldots + b_l n^{t_l}},$$

where all the letters denote positive numbers and the s's and t's are rational and arranged in descending order of magnitude.

5. Prove that, if $m > 0$, then

$$\frac{1}{m^2} = \frac{1}{(m+1)^2} + \frac{1}{(m+2)^2} + \ldots < \frac{m+1}{m^2}.$$

6. Prove that

$$\sum_1^\infty \frac{1}{n^2+1} < \frac{1}{2} + \frac{1}{4}\pi.$$

7. Prove that

$$-\frac{1}{2}\pi < \sum_1^\infty \frac{a}{a^2+n^2} < \frac{1}{2}\pi.$$

8. Prove that

$$2\sqrt{n-2} < \frac{1}{\sqrt{1}} + \frac{1}{\sqrt{2}} + \ldots + \frac{1}{\sqrt{n}} < 2\sqrt{n-1},$$

$$\frac{1}{2}\pi < \frac{1}{\sqrt[2]{1}} + \frac{1}{\sqrt[3]{2}} + \frac{1}{\sqrt[4]{3}} + \ldots < \frac{1}{2}(n+1).$$

9. If $f(n) \to l > 1$, the series $\Sigma n - \phi(n)$ converges. If $\phi(n) \to l < 1$, it diverges.

10. Prove that, if $a > 0$, $b > 0$ and $0 < s < 1$, then,

$$\psi(n) = (a+b)^{-s} + (a+2b)^{-s} + \ldots + (a+nb) - s - \frac{(a+nb)^{1-s}}{b(1-s)}$$

tends to a limit A when n ® ¥. Prove also that,

$$y(n) - y(n^{-1}) = 0\ (n^{-s-1})$$

ALTERNATING SERIES

A series, usually infinite, in which successive terms have opposite signs, as $1 - {}^1/_2 + {}^1/_4 - {}^1/_8 +. \ldots$

A series is called *alternating* if it has the form,

$$\sum_{k=1}^\infty (-1)^{K+1_{a_k}}$$

or,

$$\sum_{k=1}^{\infty}(-1)^{ka_k,}$$

where, $ak > 0$

The alternating series,

$$\sum_{k=1}^{\infty}(-1)^{ka_k,}$$

converges if:

1. $a_k > a_{k+1}$
2. $\lim_{K \to \infty} a_k = 0$

For the Fibonacci numbers:

1, 1, 2, 3, 5, 8, 13, 21, 34, ...

we have the next alternating series: F(n)–F(n+1)+F(n+2)–F(n+3)+F(n+4)–F(n+5)+F(n+6)–F(n+7) +....

Let us find the alternating sum of all Fibonacci numbers! We shall now introduce operator of the finite differences that connects the function,

$$Hy(x) = y(x+1) - y(x)$$

with the function,

$$y = f(x)$$

It is easy to verify that,

$$H = e^D - 1$$

where D is operator of differentation.

We shall now introduce operator's equation,

$$\frac{1}{1+e^D} = \sum_{k=0}^{\infty}(-1)^k e^{nD}$$

that connects the function,

$$\frac{1}{1+e^D} y(x)$$

with the series,

$$\sum_{n=0}^{\infty}(-1)^n e^{nD} y(x) = y(x) - y(x+1) + y(x+2) - y(x+3) + (x+4) - y(x+5) + ...$$

As will be seen, we usually deal with the next operator's identity for the Fibonacci and Lucas numbers:

$$H = \frac{1}{1+H}$$

and,

$$1 + H = \frac{1}{H}$$

By using this identity, we can write:

$$\frac{1}{1+e^{D}} f(n) = \frac{1}{1+\frac{1}{H}} f(n) = \frac{H}{1+H} f(n) + e^{-2D} F(n)$$

$$= F(n-2) = \sum_{k=0}^{\infty} (-1)^{k} F(n+k)$$

or,

$$\frac{1}{1+e^{D}} L(n) = \frac{1}{1+\frac{1}{H}} L(n) = \frac{H}{1+H} L(n) = e^{-2D} L(n)$$

$$= L(n-2) = \sum_{k=0}^{\infty} (-1)^{k} L(n+k)$$

Now, we have the next alternating series of the fibonacci numbers:

Alternating series	Its sum
1 – 1 + 2 – 3 + 5 – 8 + 13 – 21 + 34 – 55 +...	1
1 – 2 + 3 – 5 + 8 – 13 + 21 – 34 + 55 – 89 +...	0
2 – 3 + 5 – 8 + 13 – 21 + 34 – 55 + 89 – 144 +...	1
3 – 5 + 8 – 13 + 21 – 34 + 55 – 89 + 144 – 233 +...	1
5 – 8 + 13 – 21 + 34 – 55 + 89 – 144 + 233 – 377 +...	3
8–13 + 21 – 34 + 55 – 89 + 144 – 233 + 377 – 610 +...	5
13 – 21 + 34 – 55+89–144 + 233 – 377 + 610 – 987 +...	8
21 – 34 + 55 – 89 + 144 – 233 + 377 – 610 + 987–1047 +...	13
34 – 55 + 89 – 144 + 233 – 377 + 610 – 987 + 1047 – 2034 +...	21
55 – 89 + 144 – 233 + 377 – 610 + 987–1047 + 2034 – 3081...	34

or for lucas numbers:

Alternating series	Its sum
1 – 3 + 4 – 7 + 11–18 + 29 – 47 + 76 – 123 +...	–1
3 – 4 + 7 – 11 + 18 – 29 + 47 – 76 + 123 – 199 +...	2
4 – 7 + 11 – 18 + 29 – 47 + 76 – 123 + 199 – 322 +...	1
7 – 11 + 18 – 29 + 47 – 76 + 123 – 199 + 322 – 521 +...	3
11 – 18 + 29 – 47 + 76 – 123 +199 – 322 + 521 – 843 +...	4
18 – 29 + 47 – 76 + 123 – 199 + 322 – 521 + 843 – 1364 +...	7
29 – 47 + 76 – 123 + 199 – 322 + 521 – 843 + 1364 – 2207 +...	11
47 – 76 + 123 – 199 + 322 – 521 + 843 – 1364 + 2207 – 3571...	18
76 – 123 + 199 – 322 + 521 – 843 + 1364 – 2207 + 3571 – 5778...	29
123 – 199 + 322 – 521 + 843–1364 + 2207 – 3571 + 5778 – 9349...	47

CONDITIONS OF CONV ERGENCE OF ALTERNATING SERIES

We have seen that, in general, for a given series Σx_n, the series $\Sigma|xn|$ may not be convergent. In other words, the series Σx_n is not absolutely convergent. So, in this case, it is almost a lost case, meaning it is very hard to use the old tools developed for positive series. But, for a very special kind of series we do have a partial answer (due to Abel). This is the case for alternating series.

These are of the form,

$$\sum_{n\geq 1}(-1)n_{a_n},$$

where $a_n \geq 0$.

Example: The series,

$$\sum_{n\geq 1}\frac{(-1)^{n+1}}{n} = 1-\frac{1}{2}+\frac{1}{3}-\frac{1}{4}+\frac{1}{5}-\frac{1}{6}+\ldots$$

is alternating. Note that we do not have $(-1)^n$, but $(-1)^{n+1}$. This is not important. What matters is that the sign alternates; once it is positive, the next one is negative and so on....

Alternating Series Converge

In order to appreciate Abel's result on alternating series, let us play with the above example. Indeed, consider the series,

$$\sum_{n\geq 1}\frac{(-1)^{n+1}}{n}$$

Let us generate the sequence of partial sums $\{s_n\}$. We have,

$$s_1 = 1$$

$$s_2 = 1-\frac{1}{2} = 0.5$$

$$s_3 = 1-\frac{1}{2}+\frac{1}{3} = 0.833$$

$$s_4 = 1-\frac{1}{2}+\frac{1}{3}-\frac{1}{4} = 0.583$$

$$s_5 = 1-\frac{1}{2}+\frac{1}{3}-\frac{1}{4}+\frac{1}{5} = 0.783$$

$$s_6 = 1-\frac{1}{2}+\frac{1}{3}-\frac{1}{4}-\frac{1}{5}-\frac{1}{6} = 0.616$$

This clearly implies,

$$s_2 < s_4 < s_6 \qquad < \qquad < s_5 < s_3 < s_1.$$

So, one may wonder whether the even $\{s_{2n}\}$ sequence is increasing and the odd sequence $\{s_{2n+1}\}$ is decreasing while satisfying,

$$s_{2n} < s_{2n+1}.$$

The answer is: YES. Indeed, we have,

$$s_{2n+1} = s_{2n} + \frac{1}{2n+1},$$

which implies $s2n < s2_{n+1}$. Let us check that $\{s_{2n}\}$ is increasing (the odd one is left to the reader to prove). We have,

$$s_{2n+1} = s_{2n} + \frac{1}{2n+1}\frac{1}{2n+2}.$$

Since,

$$\frac{1}{2n+1}-\frac{1}{2n+2} = \frac{1}{(2n+1)(2n+2)} > 0,$$

then we have $s_{2n} < s_{2n+2}$, which implies that $\{s_{2n}\}$ is increasing. By the way, we did evaluate,

$$\frac{1}{2n+1}-\frac{1}{2n+2},$$

but, in fact, we did not need that. What makes it work is the fact that the sequence,

$$\left\{\frac{1}{n}\right\}$$

is decreasing. So, since the sequence $\{s_{2n}\}$ is increasing and bounded above by s_1, then it is convergent to a number A. The same holds for $\{s_{2n+1}\}$ which is decreasing and bounded below by s_2. Hence, it is convergent to a number B. We must have,

$$s_{2n} \leq A \leq B \leq s2_{n+1.}.$$

This obviously implies,

$$B-A \leq s_{2n+1}-s_{2n}-=\frac{1}{2n+1}..$$

Since,

$$\lim_{x\to\infty} s_{2n+1}-s_{2n} = \lim_{x\to\infty}\frac{1}{2n+1}=0,$$

we deduce that we must have $A = B$, which gives,

$$\lim_{x\to\infty} s_{2n+1} = \lim_{x\to\infty} s_{2n}.$$

This clearly implies that the sequence is convergent and,

$$\lim_{x\to\infty} s_n = \lim_{x\to\infty} s_{2n+1} = \lim_{x\to\infty} s_{2n}.$$

Therefore, the series,

$$\sum_{n\geq 1}\frac{(-1)^{n+1}}{n}$$

is convergent and we have,

$$\sum_{n=1}^{\infty}\frac{(-1)^{n+1}}{n} = \lim_{x\to\infty} s_{2n+1} = \lim_{x\to\infty} s_{2n}.$$

From the above inequalities, we get,

$$1-\frac{1}{2}+...-\frac{1}{2n}<\sum_{n=1}^{\infty}\frac{(-1)^{n+1}}{n}<1-\frac{1}{2}+...+\frac{1}{2n}+\frac{1}{2n+1},$$

for any $n \leq 1$. These inequalities allow for an approximation of the total sum by the partial sums. If you wonder what the total sum is, the answer is (by using Taylor series):

$$In\sum_{n=1}^{\infty}\frac{(-1)^{n+1}}{n}$$

Remark: Let us give another way to prove,

$$In(2)\sum_{n=1}^{\infty}\frac{(-1)^{n+1}}{n}.$$

First, consider the sequence $\{c_n\}$ defined by,

$$c_n = 1+\frac{1}{2}+...+\frac{1}{n}-\ln(n).$$

It is easy to check that $c_n \geq 0$, for $n \geq 1$. Let us show that is decreasing.

Indeed, we have,

$$c_{n+1}-c_n = \frac{1}{n+1}-\ln(n+1)+\ln(n) = \frac{1}{n+1}-\ln\left(\frac{n+1}{n}\right)..$$

Set,

$$f(x)=\frac{1}{x+1}-\ln\left(1+\frac{1}{x}\right) \text{ for } x \geq 1.$$

We have,

$$f'(x) = -\frac{1}{(x+1)^2}+\frac{\frac{1}{x^2}}{1+\frac{1}{x}} = \frac{1}{x(1+x)} \geq 0.$$

Hence, the function $f(x)$ is increasing for $x \in [1,\infty$.

Since,

$$\lim_{x\to\infty} f(x)=0,$$

then we must have $f(x) \leq 0$ for $x \geq 1$.

This implies,
ln,

$$\left(1+\frac{1}{x}\right) \geq \frac{1}{x+1}$$

for $x \geq 1$.

Hence, we have,

$$\left(\frac{n+1}{n}\right) \geq \frac{1}{n+1}$$

for $x \geq 1$. Therefore, the sequence $\{c_n\}$ is decreasing.

Since it is bounded below by 0, we conclude that $\{c_n\}$ is convergent.

Write,

$$\lim_{x \to \infty} 1 + \frac{1}{2} + \ldots + \frac{1}{n} - \ln(n) = C.$$

C is called Euler's constant. We have,

$$0 < C \leq + \frac{1}{2} + \ldots + \frac{1}{n} - \ln(n),$$

for any $n \geq 1$. Let us go back to the alternating series

$$\sum_{n \geq 1} \frac{(-1)n+1}{n}.$$

We have,

$$\sum_{k=1}^{k=2n} \frac{(-1)k+1}{k} = \sum_{k=1}^{k=2n} \frac{1}{k} - \sum_{k=1}^{k=n} \frac{1}{k} = c_{2n} - c_n + \ln(2).$$

Since,

$$\lim_{x \to \infty} c_{2n} - c_n + \ln(2) = \ln(2),$$

we conclude that,

$$\sum_{n=1}^{\infty} \frac{(-1)^{n+1}}{n} = \ln(2).$$

So what did we learn from the above example?

By looking carefully at the above calculations, we may be able to come up with a more general result.

Alternating Series Test

Consider the alternating series,

$$\sum_{n\geq 1}(-1)^{n_{a_n}}$$

where $a_n \leq 0$. Assume that:

1. $\{a_n\}$ is decreasing;
2. $\lim\limits_{x\to\infty} a_n = 0$;

Then the series,

$$\sum_{n\geq 1}^{\infty}(-1)^{n_{a_n}}$$

is convergent. Moreover, the estimate of the total sum,

$$\sum_{n=1}^{\infty}(-1)^{n_{a_n}},$$

by the nth partial sum , has error of magnitude at most. In other words, we have,

$$\left|\sum_{n=1}^{\infty}(-1)^{n_{a_n}} - s_n\right| \leq a_{n+1}.$$

Example: Classify the series,

$$\sum_{n\geq 1}(-1)^n \frac{\ln(n)}{n}$$

as either absolutely convergent, conditionally convergent, or divergent.

Answer: Consider the series of the absolute values,

$$\sum_{n\geq 1}\left|(-1)^n \frac{\ln(n)}{n}\right| = \sum_{n\geq 1}\frac{\ln(n)}{n}.$$

This is a Bertrand Series with $\alpha = 1$ and $\beta = 1$. Using the Bertrand Series Test, we conclude that it is divergent. Hence, the series,

$$\sum_{n\geq 1}(-1)^n \frac{\ln(n)}{n}$$

is not absolutely convergent. Since this series is alternating, with,

$$a_n = \frac{\ln(n)}{n},$$

let us check if the assumptions of the Alternating Series Test are satisfied. First, we need to check that $\{a_n\}$ is decreasing.

Set,

$$f(x) = \frac{\ln(x)}{x} \textit{ for } x \geq 1.$$

We have,

$$f'(x) = \frac{1-\ln(x)}{x^2}.$$

Clearly, we have $f'(x) < 0$, for $x > e$. Hence, the sequence,

$$\left\{\frac{\ln(n)}{n}\right\} n \geq 3$$

is decreasing. It is easy to check that,

$$\lim_{x\to\infty} \frac{in(n)}{n} = 0.$$

Therefore, all the Alternating Series Test assumptions are satisfied. We then conclude that the series,

$$\sum_{n\geq 1}(-1)^n \frac{\ln(n)}{n}$$

is convergent. In fact, in order to be precise it is conditionally convergent.

Example: Classify the series,

$$\sum_{n\geq 1} \frac{\sin(n\pi/2)}{n}$$

as either absolutely convergent, conditionally convergent, or divergent.

Answer: It is not clear from the definition what this series is. So we advise you to take your calculator and compute the first terms to check that in fact we have,

$$\sum_{n\geq 1}\frac{\sin(n\pi/2)}{n}=1-\frac{1}{3}+\frac{1}{5}-\frac{1}{7}+\frac{1}{11}+\frac{1}{13}-\frac{1}{15}+\ldots$$

This is the case because,

$$\sin(n\pi/2)=\begin{cases}0 & if \quad n \quad is \quad even\\ \pm 1 & if \quad n \quad is \quad odd\end{cases}$$

So, this is an alternating series with,

$$a_n=\frac{1}{2n+1}.$$

Since this sequence is decreasing and goes to 0 as , then by the Alternating Series Test, the series,

$$\sum_{n\geq 1}\frac{\sin(n\pi/2)}{n}-1-\frac{1}{3}+\frac{1}{5}-\frac{1}{7}+\ldots$$

is convergent. Note that it is not absolutely convergent.

Bibliography

Andrews, B.: *The Affine Curve-lengthening Flow,* NJ: Rutgers University Press, 2001.

Bokan, P.: *Geometry of Differential Operators on Weyl Manifolds,* New York: Oxford University Press, 2002.

Furuhata, H.: *Open problems in affine differential geometry and related topics,* New Delhi: Long Life Publication, 2003.

Gigena, S.: *Ordinary Differential Equations in Affine Geometry: Differential Geometric Setting and Summary of Resul,* London: Routledge, 2002.

Gromov, M.: *Sign and Geometric Meaning of Curvature,* NJ: Rutgers University Press, 2001.

Izumiya, T.: *Affine Differential Geometry of Space Curves and Singularity Theory,* Dharwar: Karnataka University Press, 2007.

Liao, H. K.: *Heat Flow Problems in Centroaffine Geometry,* New York: Oxford University Press, 2002.

Magid, C.: *Recent Developments in Affine Differential Geometry,* New York: Oxford University Press, 2002.

Malakhovsky, V. S.: *Differential Geometry of Manifol,* NJ: Rutgers University Press, 2001.

Matsuzoe, H.: *Geometry of Contrast Functions and Conformal Geometry,* New Delhi: Long Life Publication, 2003.

Olshanetskii, M. A.: *A Short Guide to Modern Geometry for Physicists,* NJ: Rutgers University Press, 2001.

Peikert, M.: *Examples of Weyl Geometries in Affine Differential Geometry,* New Delhi: Long Life Publication, 2003.

Scharlach, C.: *Centroaffine Differential Geometry of Definite Oriented Surfaces,* New York: Oxford University Press, 2002.

Simon, G.: *Global Affine Differential Geometry of Hypersurfaces,* New York: Oxford University Press, 2002.

Simon, U.: *Affine Differential Geometry,* New Delhi: Long Life Publication, 2003.

William, M.: *An Introduction to Differentiable Manifolds and Riemannian Geometry",* Orlando: Academic Press, 2005.

Yagi, B.: *Geometry of Hessian manifolds,* Dharwar: Karnataka University Press, 2007.

Index

T

U

V